Safeguarding Pipelines: The Crucial Role of Air Pocket Management in Transient Flow

Veron

Contents

3 Transient phenomena during the emptying process of a single pipe with water-air interaction   31

4 Subatmospheric pressure in a water draining pipeline with an air pocket   60

Chapter 1

Introduction

1.1 Background

The analysis of transient phenomenon in water distribution systems is important due to the intricacy of computations and the particularities of the several configurations of each system. The phenomenon is complex to analyze even for a single-phase (water). When entrapped air pockets are present into installations, the analysis is much more complex because there are two fluids (water and air) in two different phases (liquid and gas) (Fuertes, 2001).

Filling and emptying processes may introduce entrapped air pockets to pipelines, where the implementation of hydraulic (water phase) and thermodynamic (air phase) equations is required since typical transient flow models (e.g Allievi, Hammer, among others) are not capable of predicting a two-phase flow (water and air). These processes need to be implemented by the technical personnel working at water supply systems to protect hydraulic installations.

These operations are periodically performed in real hydraulic installations. The filling operation can generate dangerous pressure surges when air pockets are rapidly compressed and air valves are not expelling enough quantity of air (Izquierdo et al., 1999; Zhou et al., 2013a; Wang et al., 2017); while the emptying operation can produce subatmospheric pressure values due to the expansion of air pockets if air valves are not capable of admitting air flow with a similar ratio of the drained water (Coronado-Hernández et al., 2017b). Extreme values of the absolute pressure depend on the characteristics of installations (internal diameter, pipe roughness and pipe slope). Extreme pressures can affect not only pipelines but also their devices (joints, pumps' stations and valves).

Filling processes have been studied by several authors (Izquierdo et al., 1999; Zhou et al., 2013a; Zhou and Liu, 2013; Wang et al., 2017) based on the behavior of hydraulic and thermodynamic variables. However, it has been detected, in the literature, a lack of knowledge regarding the analysis of transient flow of a water emptying pipeline, where only semi-empirical models have been developed in this field (Tijsseling et al., 2016; Laanearu et al., 2012); in this sense, previous models cannot be extended for analyzing complex systems. A robust mathematical model based on physical equations is then required to compute the evolution of the main hydraulic and thermodynamic variables during emptying processes in pressurized pipelines of irregular profiles. These variables are: air absolute pressure, air density, water velocity, air flow, and length of a water column. At present, engineers follow some recommendations of the American Water Works Association (AWWA, 2001) and manufacturers to avoid lowest values of subatmospheric pressure, which can cause a pipeline collapse during an emptying process. Air valves are installed along of piping systems to avoid both the risk of pipelines collapse and the cavitation occurrence by admitting air into installations until atmospheric conditions are attained. Entrapped air can be entered in hydraulic installations not only by air valves but also through joints and water intakes, during pumps' stoppages or failure of the systems, by releasing of dissolved air, and by vortex formation at pump inlets.

1.2 Motivation

Subatmospheric pressure patterns can be reached in water distribution systems with transient flow occurrence under two scenarios: (i) with a single phase (water), as occurs in pump's stoppages, and by closing valves; and (ii) with a two-phase flow (air-water mixture) which occurs during an emptying process. At present, there is no mathematical model to simulate a transient flow during a water emptying process, which is crucial to ensure pipeline protection. In this sense, protection devices need to be sized not only in transient flow with a single phase (typical practice) but also during the emptying process (a missing practice). Hence, the most critical condition should be used to size a pipe stiffness class depending on the minimum value of a reached subatmospheric pressure during transient flows (both single and two-phase flow) and characteristics of pipe installations (the type of soil in natural conditions, the type of backfill and the cover depth).

Some situations are reported in the literature where an expansion of air pockets generated a systems collapse. Espert et al. (1991) reported an air vessel collapse with a volume of 56 m^3 located at Mediterranean coast, which was generated by a thermodynamic expansion of an air pocket with a reached subatmospheric pressure value of 0,22 bar. De Paor et al. (2012) mentioned an example of pressure vessel collapse due to sheet blocking vent during a vacuum condition. Tullis and Watkins (1991) studied the collapse of a large pipeline of a hydropower (with a diameter of 1,3 m and 4825 m long.) due to a pipe rupture which was generated by a low value of subatmospheric pressure, who mentioned that if an appropriate air valve had been installed, then the pipe would not have collapsed. Cabrera et al. (2008) studied different emergency scenarios related to transient flow of a 70-km-long pipeline with diameter of 1,6 m, where a total pipe rupture was analyzed highlighting the importance to select adequately air valves to prevent pipe collapse during filling and emptying processes.

This research develops a mathematical model to predict emptying operations considering different configurations of pipelines. A combination of hydraulic and thermodynamic equations is required for simulating the emptying operation. The mathematical model uses the rigid water column model, the piston flow model (air-water interface perpendicular to the pipe direction), and the polytropic model of the entrapped air pocket. When air valves are acting, then the continuity equation of the air pocket and the air valve characterization are included in the definition of the mathematical model. Two experimental facilities are used to validate the mathematical model.

1.3 Outline of this book

1.3.1 Objectives

The analysis of transient phenomena during the filling process of pipelines have been studied by several authors; however, there is a lack of detailed studies related to the emptying process of water pipelines, which is very important in order to avoid the systems collapse on the trough of subatmospheric pressure reached by air pockets. The overall objective of this research is:

To investigate as deeply as possible about the analysis of transient phenomenon of a water emptying pipeline of irregular profile with several air pockets and various air valves

The overall objective was divided into several detailed objectives:

1. To perform a literature review regarding the analysis of transient phenomena with trapped air in water distribution networks..

2. To develop a rigid water column model to simulate the emptying process of water in single pipelines for two cases: (i) with an upstream end closed, and (ii) with admitted air located at upstream end.

3. To solve a differential-algebraic system to simulate the emptying process of water in single pipelines using the Simulink Library in Matlab.

4. To prepare an experimental facility of a single pipe for simulating an emptying process for the two aforementioned cases by measuring the air absolute pressure to validate the mathematical model.

5. To develop a mathematical model for analyzing a water emptying pipeline of irregular profile for two cases: (i) without the admitting air, and (ii) using air valves.

6. To solve a set of differential-algebraic equations of emptying processes considering pipelines of irregular profiles using the Simulink Library of Matlab.

7. To validate the mathematical model of a water emptying pipeline of irregular profile through measurements of main hydraulic and thermodynamic variables (water velocity, air absolute pressure, and lengths of the water columns).

8. To perform a sensitivity analysis of the main hydraulic and thermodynamic parameters during emptying processes.

9. To compare results of the mathematical model with previous models to simulate the emptying process of water pipelines.

10. To apply the mathematical model to simulate an emptying process under different scenarios to check the risk of collapse of the pipeline network of the neighborhood Ciudad del Bicentenario located in Cartagena de Indias, Colombia.

1.3.2 Methodology

This research is developed considering the following steps:

- **A literature review:** a complete literature review regarding the analysis of transient phenomena with trapped air was conducted considering these journals: Journal of Hydraulic Research, Journal of Hydraulic Engineering, Journal of Water Resources Planning and Managment, Water MDPI Journals, Urban Water Journal, Canadian Journal of Civil Engineering, among others. Also, conference contributions were investigated.

- **Mathematical models:** to simulate a transient flow with entrapped air pockets is necessary to understand the behavior of both water and air phase. In this sense, the water phase was simulated considering a rigid water column model, which provides sufficient accuracy by taking a moving air-water interface. Moreover, the air phase was simulated with a polytropic equation, the continuity equation of air pockets, and the air valve characterization. Computational simulations were performed using the Simulink Library of Matlab.

- **Experimental facility and instrumentation:** two experimental facilities were used to validate the mathematical model. A complete description of a single pipe is presented in Chapter 3, and about a pipeline of irregular profile is shown in Chapters 4 and 5. To record the information of the main hydraulic variables, the following instrumentation was used: (i) absolute pressure pattern was measured by using different transducers located at the high point of both installations; (ii) water velocity was measured using an Ultrasonic Doppler Velocimetry (UDV) avoiding as much as possible the noise (electrical, hydraulic, vehicular, among others); (iii) and lengths of water columns were recorded with a typical Camera.

- **Verification of the mathematical model and limitations of proposed model:** to validate the mathematical model was conducted a comparison between computed and measured of the main hydraulic variables (absolute pressure pattern, water velocities, and lengths of water columns). Also, the comparison showed the need to include some limitations of the proposed model.

- **Application to a case study:** the mathematical model was applied to study the emptying process of water of a case study.

1.3.3   book **structure**

An introduction is presented in this Chapter describing a background, the motivation, the proposed objectives, and the structure of this research. The remaining chapters are clearly identified as follows:

- Chapter 2 presents a literature review about transient flow with trapped air.

- Chapter 3 describes numerical and experimental analysis of a water emptying process performed in a single pipe.

- Chapter 4 shows an experimental and numerical analysis of a water emptying pipeline of irregular profile with entrapped air.

- Chapter 5 is similar to Chapter 4, but the behavior of air valves have been included in the proposed model.

- Chapter 6 contains additional experiments to validate the proposed model.

- Chapter 7 presents an application of the proposed model to a case study.

- Chapter 8 summarizes the discussion of this research, showing how each one of the objectives was reached.

- Chapter 9 presents the main contributions, conclusions and future developments of the book.

At the end of the document is presented the references used in this research.

1.3.4   List of publications

This book is composed by collecting papers (Chapters 2 to 7) according to guidelines of the Universitat Politècnica de València, which are listed below:

1. *Hydraulic modeling during filling and emptying processes in pressurized pipelines: A literature review*

   (a) Coauthors: Fuertes-Miquel, V. S; Coronado-Hernández O. E; Mora-Meliá, D; Iglesias-Rey P. L.

   (b) Journal: Urban Water Journal ISSN 1573-062X.

   (c) 2017 Impact Factor: 2.744. Position JCR 17/90 (Q1). Water Resources.

   (d) State: Accepted with revisions. January 2019.

2. *Transient phenomena during the emptying process of a single pipe with water-air interaction*

   (a) Coauthors: Fuertes-Miquel, V. S; Coronado-Hernández O. E; Iglesias-Rey P. L.; Mora-Meliá, D.

   (b) Journal: Journal of Hydraulic Research ISSN 0022-1686

   (c) 2017 Impact Factor: 2.076. Position JCR 41/128 (Q2). Civil Engineering

   (d) State: Published. Latest Articles. 2018; DOI: 10.1080/00221686.2018.1492465

3. *Subatmospheric pressure in a water draining pipeline with an air pocket*

   (a) Coauthors: Coronado-Hernández O. E; Fuertes-Miquel, V. S; Besharat, M.; and Ramos, H. M.

   (b) Journal: Urban Water Journal ISSN 1573-062X.

   (c) 2017 Impact Factor: 2.744. Position JCR 17/90 (Q1). Water Resources.

   (d) State: Published. 2018. Volume 15; Issue 4. DOI: 10.1080/1573062X.2018.1475578

4. *Experimental and Numerical Analysis of a Water Emptying Pipeline Using Different Air Valves*

   (a) Coauthors: Coronado-Hernández O. E; Fuertes-Miquel, V. S; Besharat, M.; and Ramos, H. M.

   (b) Journal: Water ISSN 2073-4441.

   (c) 2017 Impact Factor: 2.069. Position JCR 34/90 (Q2). Water Resources.

   (d) State: Published. 2017. Volume 9; Issue 2; 98. DOI: 10.3390/w9020098.

5. *Rigid Water Column Model for Simulating the Emptying Process in a Pipeline Using Pressurized Air*

   (a) Coauthors: Coronado-Hernández O. E; Fuertes-Miquel, V. S; Iglesias-Rey, P.L.; Martínez-Solano, F.J.

   (b) Journal: Journal of Hydraulic Engineering ISSN 0733-9429.

   (c) 2017 Impact Factor: 2.080. Position JCR 39/128 (Q2). Civil Engineering.

   (d) State: Published. 2018. Volume 144; Issue 4. DOI: 10.1061/(ASCE)HY.1943-7900.0001446.

6. *Emptying Operation of Water Supply Networks*

   (a) Coauthors: Coronado-Hernández O. E; Fuertes-Miquel, V. S; Angulo-Hernández, F. N.

   (b) Journal: Water ISSN 2073-4441.

   (c) 2017 Impact Factor: 2.069. Position JCR 34/90 (Q2). Water Resources.

   (d) State: Published. 2018. Volume 10; Issue 1; 22. DOI: 10.3390/w10010022.

In addition, 6 conference presentations were conducted as presenting author:

1. *The analysis of transient flow during the emptying process in a single pipeline*

   (a) Coauthors: Oscar E. Coronado-Hernández, Vicente S. Fuertes-Miquel, Pedro L. Iglesias-Rey, and Daniel Mora-Meliá.

   (b) Event: XXVII Congreso Latinoamericano de Hidráulica, IAHR.

   (c) Year: 2016.

   (d) Place: Lima, Perú.

2. *Sensitivity analysis of emptying processes in water supply networks*

   (a) Coauthors: Oscar E. Coronado-Hernández, Vicente S. Fuertes-Miquel, Pedro L. Iglesias-Rey, and Daniel Mora-Meliá.

   (b) Event: V Jornadas de Ingeniería del Agua.

   (c) Year: 2017.

   (d) Place: A Coruña, Spain.

3. *Behavior of a water-draining pipeline of irregular profile with air valves*

   (a) Coauthors: Vicente S. Fuertes-Miquel, Oscar E. Coronado-Hernández, Pedro L. Iglesias-Rey, and F. Javier Martínez-Solano.

   (b) Event: XV Seminario Iberoamericano de Redes de Agua y Drenaje, SEREA2017.

   (c) Year: 2017.

   (d) Place: Bogotá, Colombia.

4. *Analysis of the draining operation in water supply networks. Application to a water pipeline in Cartagena de Indias, Colombia*

   (a) Coauthors: Oscar E. Coronado-Hernández, Vicente S. Fuertes-Miquel, and Fredy N. Angulo.

   (b) Event: XV Seminario Iberoamericano de Redes de Agua y Drenaje, SEREA2017.

   (c) Year: 2017.

   (d) Place: Bogotá, Colombia.

5. *Negative pressure occurrence during the draining operation of water pipelines with entrapped air*

   (a) Coauthors: Vicente S. Fuertes-Miquel, Oscar E. Coronado-Hernández, Daniel Mora-Meliá, and Pedro L. Iglesias-Rey.

   (b) Event: XV Seminario Iberoamericano de Redes de Agua y Drenaje, SEREA2017.

    (c) Year: 2017.

    (d) Place: Bogotá, Colombia.

6. *A parametric sensitivity analysis of numerically modelled piston-type filling and emptying of an inclined pipeline with an air valve*

    (a) Coauthors: Oscar E. Coronado-Hernández, Vicente S. Fuertes-Miquel, Mohsen Besharat, and Helena M. Ramos.

    (b) Event: 13th International Conference on Pressure Surges..

    (c) Year: 2018.

    (d) Place: Bordeaux, France.

Chapter 2

Hydraulic modeling during filling and emptying processes in pressurized pipelines: A literature review

This chapter is extracted from the paper:

Hydraulic modeling during filling and emptying processes in pressurized pipelines: A literature review
Coauthors: Fuertes-Miquel, V. S; Coronado-Hernández O. E; Mora-Meliá, D; Iglesias-Rey P. L.
Journal: Urban Water Journal ISSN 1573-062X.
2017 Impact Factor: 2.744. Position JCR 17/90 (Q1). Water Resources.
State: Accepted with revisions. January 2019.

2.2   Introduction

The analysis of transient phenomena of a single phase (only water) is complex considering the
intricacy of calculations and configurations of water pipelines systems. Filling and emptying
processes are too complex to be captured by single-phase models because there are two fluids
(water and air) in two phases (liquid and gas)(Fuertes, 2001). These operations must be performed periodically in pipelines for maintenance, cleaning or repairs by technical personnel
(Fuertes-Miquel et al., 2018b). Planning water distribution systems requires considering the
effects of entrapped air pockets to guarantee successful maintenance and repair procedures.
During filling processes, air pockets can be rapidly compressed, producing pressure surges;
during emptying processes, air pockets expand, causing subatmospheric conditions. As a
consequence, the modeling of the air phase is crucial to determine extreme values of absolute
pressure. The correct identification of a type of model (adiabatic, polytropic or isothermal)
and an air valve characterization have to be considered.

The effects of trapped air in water pipelines are generated basically with regards to two
features: (i) air density is much lower than water density by a ratio approximately $1 : 800$
times considering atmospheric conditions and a temperature of $20^{\circ}C$; and (ii) the elasticity of
air is much higher than the elasticity of water. The elasticity depends on the type of process
(isothermal, polytropic or adiabatic) and the absolute pressure of the air pocket. For instance,
the bulk modulus in an isothermal process, at atmospheric condition, presents a ratio of
20000 times, and for an absolute pressure of 10 bar, the ratio is of 2000 times. Some of the
problems cause by entrapped air pockets are: (i) additional head losses by increasing the
water velocity as a consequence of the reduction of the cross section (Stephenson, 1997),(ii)
pressure surges for starting or stopping the system because of the compression of the air
pocket, (iii) reduction of the efficiency of the pumps, (iv) vibrations in pipelines, which
generate rapid changes in the water velocity pattern, (iv) pipe corrosion owing to temperature
and absolute pressure changes, and (v) troughs of subatmospheric pressure by the expansion
of the air pocket (Fuertes-Miquel et al., 2018b; Coronado-Hernández et al., 2018a).

Entrapped air pockets are a problem for pipeline operations. It is important to perform a
filling process correctly to eliminate air pockets from hydraulic systems. However, entrapped
air could also enter through air valves, joints and valves, during the stopping and failure of
hydraulic systems, during the release of dissolved air, and by vortex generation at pump inlets.
High points along pipelines are likely locations for the accumulation of air pockets (AWWA,

2001; Ramezani and Karney, 2017), which can experience pressure surges during a filling process (Zhou et al., 2013b; Fontana et al., 2016; Martins et al., 2016; Covas et al., 2010) or drops of subatmospheric pressure during an emptying process (Fuertes-Miquel et al., 2018b; Tijsseling et al., 2016; Coronado-Hernández et al., 2018d). Air valves are used as protective devices to avoid these situations (AWWA, 2001).

Entrapped air can be expelled by permanent joints of hydraulic systems to the atmosphere (downstream conditions, fire hydrants, among others), but the most common method is via air valves located along the system (Martino et al., 2001), such as high points or others where entrapped air can accumulate.

The emptying and filling of a pipeline cannot be simulated with the commonly used 1D transient commercial packages like Bentley Hammer, H2O Surge, Allievi, among others; since they are not capable of predicting a two-phase flow (water and air).

This paper aims to accomplish the following: (1) review available knowledge regarding filling and emptying processes in pressurized hydraulic systems, which is missing in the current literature; (2) describe mathematical models for water and air phase; (3) describe methods of resolutions of transient flow during pipeline operations; (4) mention the sources of uncertainty in current models; and (5) comment regarding some considerations to protect pipelines during these operations.

2.3   Understanding of filling and emptying processes

A basic hydraulic scheme is presented to show the performance of filling and emptying processes (Figure 2.1).

The filling process begins when a discharge valve is opened (K_u). Consequently, the water column driven by an energy source (tank or pump) starts to fill the hydraulic system (see Figure 2.1a) by compressing the entrapped air pocket (Zhou et al., 2013a; Martins et al., 2017; Malekpour et al., 2016) producing peaks of absolute pressure (pressure surges). Valve K_d is closed to produce the rapid compression of an air pocket. The process is adequately completed when the entrapped air is replaced by the water column (Apollonio et al., 2016; Balacco et al., 2015). Two situations can occur: (i) when the air valve (AV) has been installed, the process finishes successfully with a well-sized device; and (ii) when there is no air valve, it results in an unfinished process and the highest pressure surges.

To note the order of magnitude of a filling process, a numerical analysis has been conducted in a single pipeline where the authors used an own code in Matlab (Fuertes-Miquel et al., 2016, 2018b; Coronado-Hernández et al., 2017b, 2018d). Figure 2.2 shows the peaks of the absolute pressure for the two aforementioned situations using the following data: $L_t = 600$ m, $D = 0,3$ m, $f = 0,018$, $n = 1,2$, $D_{av} = 50$ mm, $C_{exp} = 0,6$, $\Delta z = 12$ m, $x_0 = 500$ m, and $K_u = 0,45$ m/(m^{3}s^{-1})^2. A comparison is conducted with a relative value of air pocket pressure (p_a^*/p_{atm}^*) and the time. According to the results, when an air valve has not been installed, the maximum peak of the absolute pressure head is rapidly reached at 112,1 s with a value of 2,97 p_a^*/p_{atm}^*. After that, oscillations of the absolute pressure continue. However, when an air valve is working, the maximum value is 1,29p_a^*/p_{atm}^* (at 53,4 s).

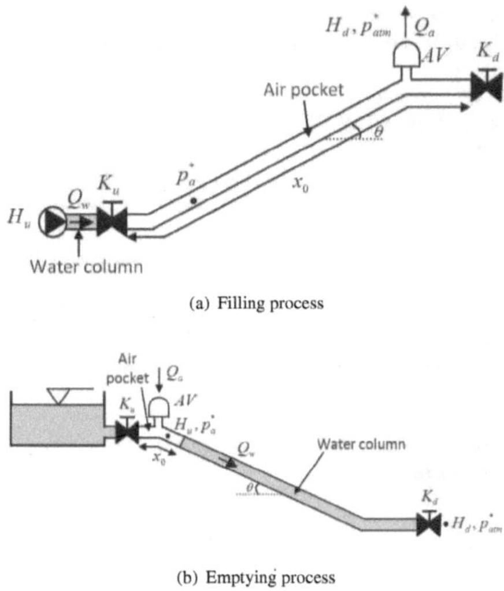

(a) Filling process

(b) Emptying process

Figure 2.1: Basic hydraulic scheme

A reduction of an absolute pressure of $1{,}68p_a^*/p_{atm}^*$ upon the peak value is reached by using a protection device (air valve) and the pipeline is completely filled at $113{,}4$ s. A pipeline with no air valve implies an incomplete filling and an air pocket trapped inside the installation. Maximum values of absolute pressure for different scenarios (such as filling operation, pump failure, among others) should be compared to select an appropriate pipe resistance.

During an emptying operation, water flow will be replaced by air flow (Tijsseling et al., 2016; Fuertes-Miquel et al., 2018b; Coronado-Hernández et al., 2018c). Figure 2.1b presents a scheme of the operation. The operation starts when drain valves located at the low points of the pipeline are opened (K_d). Then, the entrapped air pockets start to expand, generating troughs of the subatmospheric pressure, which can cause the system to collapse (Coronado-Hernández et al., 2017b). A single pipeline was considered with data similar to that of the filling process, but using $K_d = K_u$, $C_{adm} = C_{exp}$, and $x_0 = 100$ m. An entrapped air pocket is assumed to be at atmospheric conditions. Figure 2.3 shows the results regarding the relative absolute pressure. According to the results, when there is no air valve, the trough of the absolute pressure head is $0{,}3p_a^*/p_{atm}^*$ (at $145{,}8$ s), which continues until the end of the hydraulic event, and part of the water column remains inside the single pipeline. By

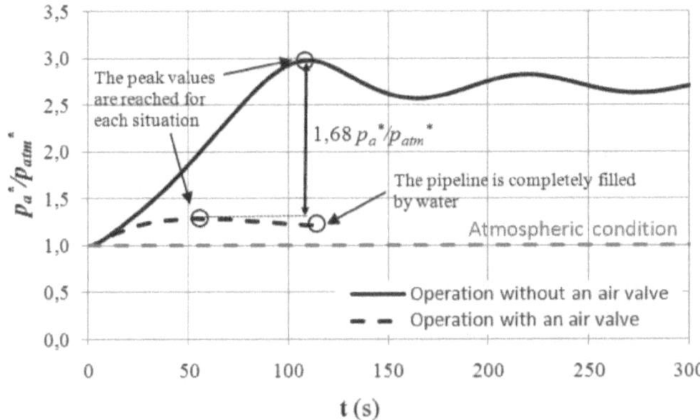

Figure 2.2: Evolution of the absolute pressure head during the filling process for different conditions

contrast, when an air valve is installed, the relative absolute pressure is $0,91p_a^*/p_{atm}^*$ (at 77,2 s), reducing the trough of relative absolute pressure by $0,61p_a^*/p_{atm}^*$ compared to the condition without an air valve. Under this situation, the hydraulic event finishes around 281,0 s, and the water column is completely drained.

2.4 Review of mathematical models

Mathematical models are complex to develop when there is water column separation (Bergant et al., 2006). Water column separation can occur for different reasons such as (i) cavitation (Yang, 2001), (ii) when the water column movement finds an entrapped air pocket, as occurs during filling and emptying processes (Martins et al., 2017), or (iii) for releasing of the dissolved air in saturated water (Ramezani et al., 2016). Pipelines always carry an air-water mixture during filling and emptying operations. Depending on the flow characteristics and conditions of pipelines, these operations can be modeled using a piston-flow model (Cabrera et al., 1992) or two-phase flow model (Bousso et al., 2013). Assumptions to apply these models depend on water velocity, internal diameter, and hydraulic slope.

The piston-flow model can be used when a hydraulic event is fast. In this case, the air-water interface is perpendicular to the pipe direction (see Figure 2.4) (Fuertes, 2001; Lee, 2005). As a consequence, along the pipeline, there are pipe branches completely filled with water and the remaining by air. The smaller the internal diameter is, the higher the water

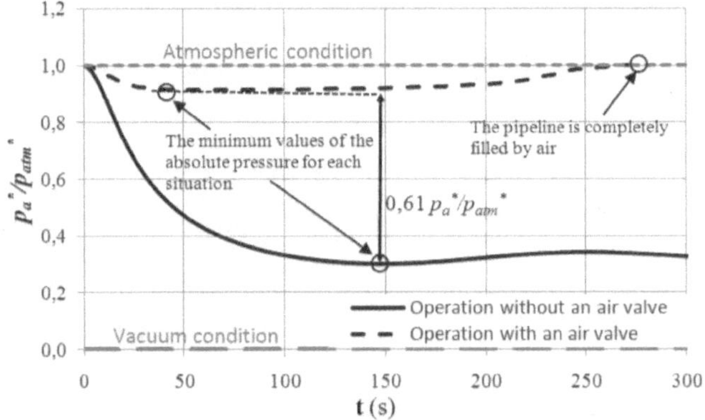

Figure 2.3: Evolution of the absolute pressure head during he emptying process for
different conditions

velocity and pipe slope. The piston-flow model is used by the majority of mathematical
models for simulating filling and emptying processes in pipelines (Liou and Hunt, 1996; Zhou
et al., 2013b; Coronado-Hernández et al., 2017b; Fuertes-Miquel et al., 2016). The length of
a water column (L) and its water velocity (v_w) are related by the following formulation:
$dL/dt = \pm v_w$ (Izquierdo et al., 1999; Zhou et al., 2013a).

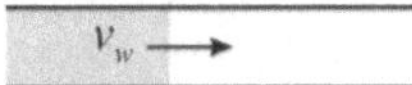

Figure 2.4: Piston-flow model

Two-phase flow models (see Figure 2.5) can be used to analyze slow transient flows and
a small quantity of air volume. The cavitation occurrence is an application of these models in
pressurized systems. These models are classified as bubble flow, bubble and air pocket flow,
plug flow, stratified wave flow, and stratified smooth flow (Bousso et al., 2013). Vasconcelos
and Wright (2008); Vasconcelos and Marwell (2011); Guinot (2001) used two-phase flow
models to understand the behavior of transient flow with trapped air in free-surface flow.

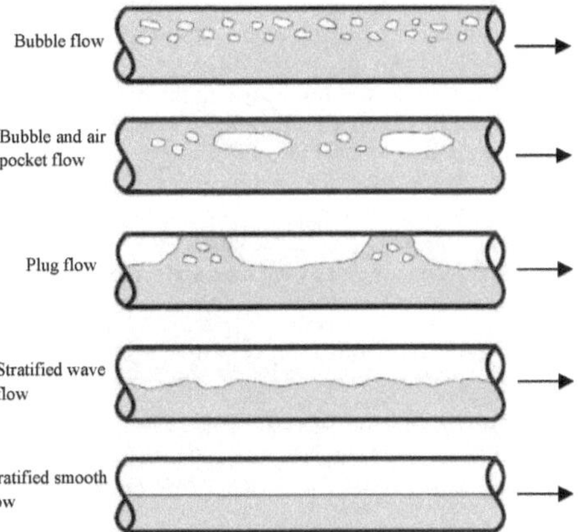

Figure 2.5: Two-phases flow models

2.4.1 Water phase

Inertial models can be used to simulate transient phenomena (Abreu et al., 1999) such as filling and emptying processes in pressurized pipelines. Inertial models consider system inertia.

There are two types of inertial models (Zhou et al., 2011b): (i) water hammer or elastic models, which consider the elasticity of the pipe and the water; and (ii) mass oscillation or rigid models, which neglect these factors.

The elastic model can be written in the simplified form as follows (Chaudhry, 2014; Wylie and Streeter, 1993):

- Mass conservation equation

$$\frac{gA}{a^2}\frac{\partial H}{\partial t} + \frac{\partial Q_w}{\partial X} = 0 \tag{2.1}$$

- Momentum equation

$$\frac{\partial Q_w}{\partial t} + gA\frac{\partial H}{\partial X} + f\frac{Q_w|Q_w|}{2DA} = 0 \tag{2.2}$$

where g = gravity acceleration, A = the cross-sectional area of the pipe, t = time, a = wave speed, X = the distance along a pipe, H = the piezometric head, Q_w = water discharge, f = the friction factor, and D = the internal diameter.

Elastic models have been used for analyzing the filling processes in water pipelines and analyzing the influence of entrapped air pockets (Zhou et al., 2011a), the effects of two entrapped air pockets (Zhou et al., 2013a), the phenomenon of white mist with entrapped air pockets (Zhou et al., 2013b), and the consequences of using a bypass (Wang et al., 2017). However, the elasticity of an entrapped air pocket into a pipeline is much higher than the elasticity of the water and pipe. As a consequence, $a \longrightarrow \infty$ or $\partial H / \partial t \longrightarrow 0$, which implies that the water phase can be modeled by the rigid water model (RWM). The mass conservation equation reduces to:

$$\frac{\partial Q_w}{\partial X} = 0 \longrightarrow Q_w = Q_w(t) \tag{2.3}$$

Then, the momentum equation can be expressed as:

$$H_u = H_d + \frac{fL}{2gDA^2}Q_w|Q_w| + R_v Q_w|Q_w| + \frac{L}{gA}\frac{dQ_w}{dt} \tag{2.4}$$

where H_u = the upstream piezometric head, H_d = the downstream piezometric head, and R_v = the resistance coefficient of the valve.

Several investigations have been conducted using rigid models for filling processes. Izquierdo et al. (1999) and Liou and Hunt (1996) applied the rigid model for analyzing a water pipeline with different air pockets, Zhou et al. (2002) presented the analysis of the transient flow in a rapidly filling horizontal pipe with an entrapped air pocket, and Hou et al. (2014) investigated a large-scale pipeline. Regarding the emptying process, Laanearu et al. (2012) experimentally analyzed this operation in a large-scale pipeline by pressurized air, Tijsseling et al. (2016) proposed a semiempirical model to predict the emptying process by pressurized air, Fuertes-Miquel et al. (2018b) proposed and validated a mathematical model of a single pipe, and Coronado-Hernández et al. (2017b) studied emptying processes using various air valves in a water pipeline. CFD models have been established to simulate filling and emptying processes (Zhou et al., 2011b; Martins et al., 2017). Table 2.1 presents the main advantages and disadvantages of various mathematical models for the water phase during filling and emptying processes.

Table 2.1: Mathematical models for the water phase

Model	Advantages	Disadvantages
Rigid column models	Models give similar results compared to the EWM because the elasticity of air pockets is much higher than pipe and water elasticity (Coronado-Hernández et al., 2017b, 2018d) Filling and emptying processes with air valves have been developed by Fuertes-Miquel et al. (2016) and Coronado-Hernández et al. (2017b), respectively. Numerical solutions and others methods are used. Models are simpler to implement. Models require a low computing time.	Generally, mathematical models use a piston flow model to define an air-water interface; but more complex shapes can be used with additional efforts (Tijsseling et al., 2016; Laanearu et al., 2012). Models neglect pipe and water elasticity (Zhou et al., 2002; Izquierdo et al., 1999). The evolution of the backflow air phenomenon has not been implemented.
Elastic models	EWMs consider pipe and water elasticity (Zhou et al., 2011b), and formulations include water hammer effects. Numerical resolution is conducted using the MOC in combination with others methods (Zhou et al., 2013a,b; Zhou and Liu, 2013). Models are more complex to implement than the RCM. Computing times are higher compared to the RCM.	A piston flow model is used to simulate an air-water interface; however, more complex shapes of air-water interfaces have not been defined. Mathematical models of filling and emptying processes with air valves have not been defined in the literature. The evolution of the backflow air phenomenon has not been implemented.
2D/3D models	Complex shapes of air-water interface can be considered (Martins et al., 2017), and the position of air-water interfaces can be modeled adequately. Backflow air phenomenon can be simulated.	Length and time scales are not defined in pipelines using air valves during filling and emptying processes. Models require a low time step which implies a high computing time (Martins et al., 2017; Zhou et al., 2011b).

2.4.2 Air phase

During the filling and the emptying operations, alterations occur on the thermodynamic proprieties of the entrapped air pockets (Martins et al., 2015), which can produce important
changes in their absolute pressure and volume. These alterations are generated for two reasons: (i) the compression or the expansion of the air pockets, and (ii) the expelled or the
admitted air flow by air valves.

Compression and expansion of the entrapped air pocket

The compression of the entrapped air pocket occurs during the filling process. It starts with
a value of absolute pressure $p_{a,1}$, and then the initial air volume $V_{a,1}$ begins to compress,
generating a higher absolute pressure value $p_{a,2}$. For an adiabatic process ($n = 1,4$) the
air volume size is higher than for an isotherm process ($n = 1,0$), which implies that an
isotherm process is more risky for this operation because higher absolute pressure values can
be reached (see Figure 2.6a). In contrast, during the emptying process, the expansion of the
entrapped air pocket occurs. The entrapped air pocket starts with an initial air volume $V_{a,1}$
at subatmospheric pressure value $p_{a,1}$ and finishes with a higher air volume size $V_{a,2}$ (see
Figure 2.6b). An adiabatic process produces higher troughs of the subatmospheric pressure
than an isotherm process. If the hydraulic event occurs slowly enough then the process can be
considered isothermal, whereas if the hydraulic event occurs fast enough, then the process is
adiabatic because there is not enough time to produce the heat transfer in adiabatic processes.
Intermediate processes are reached in actual installations.

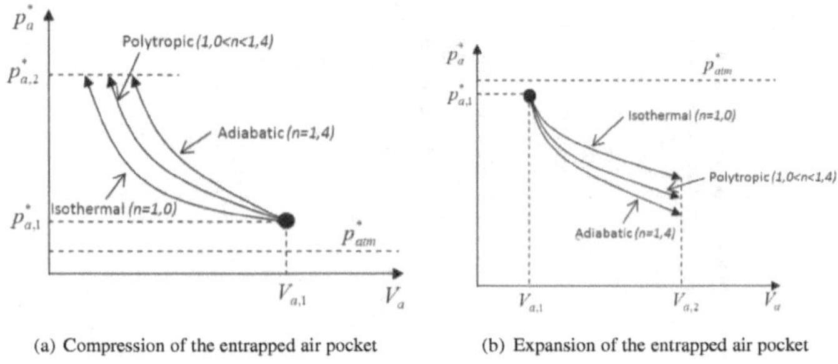

(a) Compression of the entrapped air pocket   (b) Expansion of the entrapped air pocket

Figure 2.6: Diagram p_a^* vs. V_a

An entrapped air pocket can be modeled considering its energetic behavior. According
to the first law of thermodynamics, the change in internal energy (E) is the sum of the net
quantity of heat (Q) supplied to the system plus the work done by the system (W) (Graze

et al., 1996). A simplification of these relationships for an entrapped air pocket can be represented by a polytropic model (Martin, 1976; Leon et al., 2010) where a constant polytropic coefficient (n) considers the effects of the heat transfers in the variables p_a^* and V_a. As a consequence, two formulations can be used for representing the behavior of an entrapped air pocket, which depends on whether the air mass is changing with time:

- If $dm_a/dt \neq 0$, then:

$$\frac{dp_a^*}{dt} = -n\frac{p_a^*}{V_a}\frac{dV_a}{dt} + \frac{p_a^*}{V_a}\frac{n}{\rho_a}\frac{dm_a}{dt} \tag{2.5}$$

- If $dm_a/dt = 0$, then:

$$p_a^* V_a^n = p_{a,0}^* V_{a,0}^n \tag{2.6}$$

where ρ_a is the air density, m_a is the air mass, and subscript 0 refers to initial conditions of absolute pressure of air pocket and air volume.

Air valves characterization

Air valves can be used to avoid the peaks and troughs of the absolute pressure (AWWA, 2001) for filling and emptying operations, respectively. For air valves the admitted and expelled flow air for each differential pressure should be known (Carlos et al., 2011) in order to mitigate the extreme absolute pressures. When the filling process starts ($p_a^* > p_{atm}^*$), the air valve begins to expel air out of the system and the water column starts to compress ($\Delta Q_w > 0$) the entrapped air pocket, which implies $\Delta W < 0$. In contrast, during the emptying process ($p_a^* < p_{atm}^*$), the work done by the system is $\Delta W > 0$ considering the expansion of the entrapped air pocket. Figure 2.7 shows the air valves behavior during these operations.

An appropriate modeling of air valves is crucial to understand the behavior during the filling and emptying operations in pipelines. Some theoretical expressions for modeling air valves are presented by Wylie and Streeter (1993) and Chaudhry (2014). Air valves can be simulated considering an isentropic nozzle flow that is practically adiabatic given that the time required for the air flow through this device is very small. Consequently, heat transfer cannot occur (Wylie and Streeter, 1993). The air can be modeled by applying the ideal gas law ($p_a^* = \rho_a RT$), and an adiabatic coefficient value (n) of 1,4. The air flow can be classified considering the relationships between air velocity (v_a) and the speed of sound (c). In this sense, a subsonic flow occurs if $v_a < c$, and a critical flow occurs when $v_a = c$. The speed of sound is computed as $c = \sqrt{KRT}$, where K is the polytropic coefficient in adiabatic conditions ($K = 1,4$ for air). Considering an isentropic flow, the air valves formulations for expelling and admitting air are presented as follows:

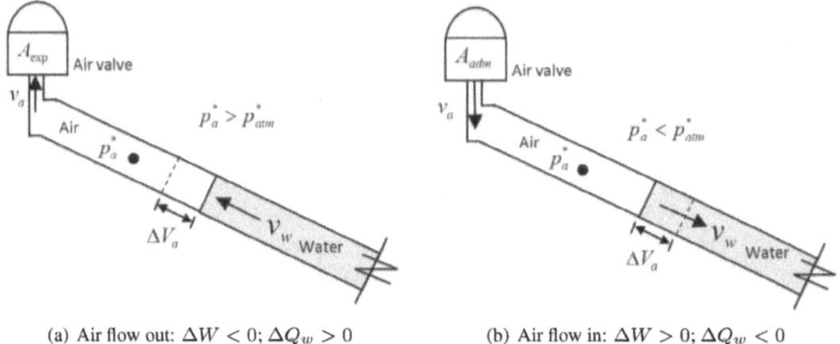

(a) Air flow out: $\Delta W < 0$; $\Delta Q_w > 0$ (b) Air flow in: $\Delta W > 0$; $\Delta Q_w < 0$

Figure 2.7: Effects of air valves behavior during the filling and emptying processes

For expelling air

- Subsonic air flow out ($p_{atm}^* < p_a^* < 1.893p_{atm}^*$):

$$Q_a = C_{exp}A_{exp}p_a^*\sqrt{\frac{7}{RT}\left[\left(\frac{p_{atm}^*}{p_a^*}\right)^{1.4286} - \left(\frac{p_{atm}^*}{p_a^*}\right)^{1.714}\right]} \qquad (2.7)$$

- Critical air flow out ($p_a^* \geq 1.893p_{atm}^*$):

$$Q_a = C_{exp}A_{exp}\frac{0.686}{\sqrt{RT}}p_a^* \qquad (2.8)$$

where C_{exp} is the outflow discharge coefficient, and A_{exp} is the cross-sectional area of the
air valve when air is expelled.

For admitting air

- Subsonic air flow in ($p_{atm}^* > p_a^* > 0.528p_{atm}^*$):

$$Q_a = C_{adm}A_{adm}\sqrt{7p_{atm}^*\rho_{a,nc}\left[\left(\frac{p_a^*}{p_{atm}^*}\right)^{1.4286} - \left(\frac{p_a^*}{p_{atm}^*}\right)^{1.714}\right]} \qquad (2.9)$$

- Critical air flow in ($p_a^* \leq 0.528p_{atm}^*$):

$$Q_a = C_{adm}A_{adm}\frac{0.686}{\sqrt{RT}}p_{atm}^* \qquad (2.10)$$

where A_{adm} is the cross-sectional area of the air valve when air is admitted, and C_{adm} is the inflow discharge coefficient.

Figure 2.8 shows a typical air valve characterization presented by manufacturers, which is obtained experimentally. The horizontal axis represents the air flow rate in normal conditions (at atmospheric pressure and ambient air temperature). The vertical axis is the differential pressure (Δp), which is computed as $\Delta p = p_a^* - p_{atm}^*$ for expelling air (filling operation) and $\Delta p = p_{atm}^* - p_a^*$ for admitting air (emptying operation).

To apply the equations 2.7 - 2.10, the inflow discharge coefficient (C_{exp}) and the outflow discharge coefficient (C_{adm}) should be obtained based on manufacturer data. In this sense, discharge coefficients (C_{exp} and C_{adm}) depend on the type of selected air valve with mean values ranging from 0,3 to 0,7 (Iglesias-Rey et al., 2014).

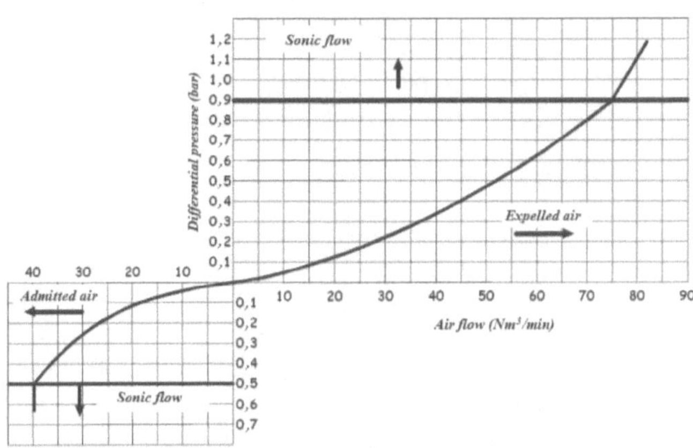

Figure 2.8: Example of an air valve characterization

2.4.3 Summary of mathematical models and methods of resolution

Table 2.2 contains a summary of mathematical models used for water, as well as the air phase, during filling and emptying processes.

Table 2.2: Summary of mathematical models for water and air phase.

Author	RCM	EWC	CFD	PF	CS	AP	AV
Coronado-Hernández et al. (2017b, 2018d)	✓			✓		✓	✓
Liu et al. (2018)		✓		✓		✓	
Wang et al. (2017)		✓		✓		✓	
Martins et al. (2017); Zhou et al. (2018b)			✓		✓	✓	
Fuertes-Miquel et al. (2016)	✓			✓		✓	✓
Tijsseling et al. (2016); Laanearu et al. (2012)	✓				✓	✓	
Hou et al. (2014)	✓			✓		✓	
Zhou et al. (2013a,b); Liu et al. (2011); Zhou et al. (2018a)		✓		✓		✓	✓
Hou et al. (2012)	✓			✓		✓	
Zhou et al. (2011b)			✓		✓	✓	
Chaudhry and Reddy (2011)	✓			✓		✓	
Malekpour and Karney (2011)		✓		✓		✓	
Lee (2005)	✓					✓	
Zhou et al. (2002)	✓			✓		✓	
Fuertes (2001); Izquierdo et al. (1999)	✓			✓		✓	✓
Liou and Hunt (1996)	✓						
Martin (1976)	✓			✓		✓	

Notes: RCM = rigid column model, EWC = elastic water column (1D), CFD = corresponds to 2D/3D CFD models, PF = air-water interface as piston flow, CS = complex shapes to represent an air-water interface with a two-phase model, AP = entrapped air pocket, and AV = air valve or orifice size.

It is important to mention how is the resolution of mathematical models for these operations. Currently, there is no single mathematical expression to directly solve the differential-algebraic equations (DAE) system for simulating the transient phenomena. Many techniques have been applied to solve a DAE system. Some of them are (i) the method of characteristics (Chaudhry, 2014; Wylie and Streeter, 1993), (ii) finite-difference methods (Cunge and Wegner, 1964), (iii) finite-element methods (Watt, 1975; Baker, 1983), (iv) finite volume method (Guinot, 2003), (v) volume of fluids (Zhou et al., 2011b; Martins et al., 2017), and (vi) smoothed particle hydrodynamics (Hou et al., 2012).

Since the 60s, the method of characteristics was used to solve the characteristic equations of a transient flow. At present, many researchers use it to model filling and emptying processes (Shimada et al., 2008; McInnis et al., 1997; Ghidaou and Karney, 1994; Wang et al., 2017; Zhou et al., 2013a) because of its computationally efficient and explicit resolution scheme.

Finite-difference methods (explicit and implicit) (Chaudhry, 2014) are useful in combination with the method of characteristics to solve complex situations in transient flow (Wang et al., 2017; Zhou and Liu, 2013). Zhou et al. (2013a,b) used this method to simulate an air-water interface and the method of characteristics to model the water phase.

Finite-element methods cannot efficiently solve problems related to the evolution of the hydraulic variables because temporal and spatial scaling are complex to model, resulting in inadequate wave propagation (Chaudhry, 2014).

Finite volume methods numerically solve the partial differential equations in the form of algebraic equations similar to finite-difference methods or finite-element methods. They are used to solve the conservation laws of hydraulic systems.

In this regard, Hou et al. (2012) presented a method based on Smoothed Particle Hydrodynamics (SPH), Zhou et al. (2011b) solved the transient flow problem using a method of Volume of Fluid (VOF), Coronado-Hernández et al. (2017b) and Izquierdo et al. (1999) used numerical solutions based on Runge-Kutta or Rosenbrock formula.

Air valves can be modeled using the method of characteristics (Ramezani et al., 2015; Zhou et al., 2013a,b) or numerical solutions of ordinary differential equations (Fuertes, 2001).

Table 2.3 presents a list of methods of resolution, where authors are mentioned. All methods of resolution are performed considering a constant friction factor, a non-variable polytropic coefficient, considering a known air pocket size, and the characteristics of air valves describe by manufacturers.

2.5 Uncertainty of current models and prospects

The sources of uncertainty that current models do not consider are the friction factor, the polytropic coefficient, the air pocket size, and the air valve behavior. These parameters can affect the determination of extreme pressures (maximum and minimum) depending on pipeline characteristics

Table 2.3: Summary of methods of resolution.

Author	NS	MOC	FD	VOF	SPH
Coronado-Hernández et al. (2017b, 2018d)	✓				
Liu et al. (2018)		✓			
Wang et al. (2017)		✓			
Martins et al. (2017); Zhou et al. (2018b)				✓	
Fuertes-Miquel et al. (2016)	✓				
Tijsseling et al. (2016); Laanearu et al. (2012)	✓				
Hou et al. (2014)	✓				
Zhou et al. (2013a,b); Liu et al. (2011); Zhou et al. (2018a)		✓	✓		
Hou et al. (2012)					✓
Zhou et al. (2011b)				✓	
Chaudhry and Reddy (2011)	✓	✓			
Malekpour and Karney (2011)		✓			
Lee (2005)			✓		
Zhou et al. (2002)	✓				
Fuertes (2001); Izquierdo et al. (1999)	✓				
Liou and Hunt (1996)	✓				
Martin (1976)	✓				

Notes: NS = numerical solutions based on Runge-Kutta or Rosenbrock formula, MOC = method of
characteristics, FD = finite-difference, VOF = volume of fluids, and SPH = smoothed particle hydrodynamics

2.5.1 Friction factor

The friction factor changes during filling and emptying processes because water velocities
vary owing to hydraulic events. Current models consider a constant friction factor during
the transient flow (Izquierdo et al., 1999; Laanearu et al., 2012; Coronado-Hernández et al.,
2017b; Zhou and Liu, 2013; Liou and Hunt, 1996; Wang et al., 2017). Some expressions
have been developed to consider a variable friction factor. Fuertes (2001) demonstrated that
the expression proposed by Brunone et al. (1991) can adequately fit the behavior of transient
flow during these operations, expressed as:

$$J_u = J_s + \frac{k_3}{g}\left(\frac{\partial v}{\partial t} - \frac{\partial v}{\partial s}\right) \tag{2.11}$$

When a rigid water model is used, the convective acceleration is null, and then:

$$J_u = J_s + \frac{k_3}{g}\frac{dv}{dt} \tag{2.12}$$

A comparison between a constant and unsteady friction factor was conducted by Fuertes
(2001) showing similar discrepancies of 1,31% (constant friction factor) and 1,05% (unsteady
friction factor) between experiments and mathematical models as shown in Figure 2.9, during the filling operation in a metacrylate pipeline of internal diameter of 18,8 mm and a total
length of 8,62 m. A similar analysis was conducted by Wang et al. (2018), where different

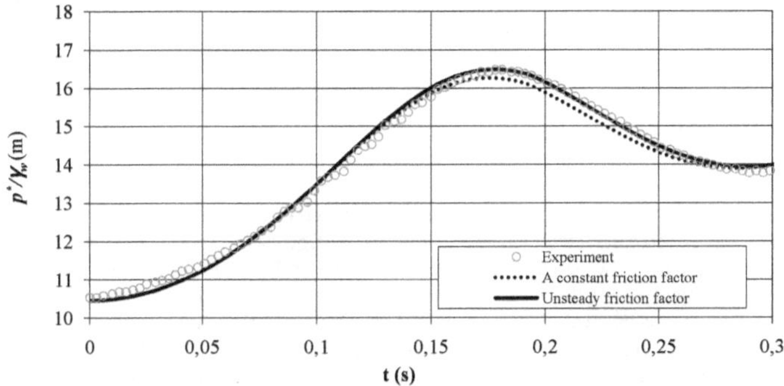

Figure 2.9: Comparison between a constant and unsteady friction factor (Fuertes, 2001)

models are presented to simulate an unsteady friction factor. Results show that both constant and unsteady friction factor can be used to simulate the filling operation; however, an unsteady friction factor improves results compared to a constant friction factor. This kind of analysis has not been performed in the emptying operation.

2.5.2 Polytropic coefficient

Current models consider the polytropic coefficient during filling and emptying processes as constant. However, it can change during a hydraulic event, depending on the installation characteristics and the boundary conditions. In this sense, a transient flow can start with an isothermal condition ($n = 1,0$), but over time, an intermediate or adiabatic ($n = 1,4$) condition can be reached, and vice versa. Fuertes-Miquel et al. (2016) shows how the selection of a polytropic coefficient in isothermal ($n = 1,0$) or adiabatic ($n = 1,4$) conditions can induce significant differences of the maximum absolute pressure attained. Many studies use an intermediate polytropic coefficient of $n = 1,2$ considering the uncertainty of this value. The determination of a reliable polytropic coefficient is important during these operations since important differences can be achieved. A filling process was analyzed in a 750-m-long pipeline, where the peak of the absolute pressure head was 113 m, which is higher compared to the peak under adiabatic conditions of 92,5 m. Here, a difference around of 18% was found.

A variable polytropic coefficient represents a challenge in current models (Wang et al.,

2018; Izquierdo et al., 1999; Zhou et al., 2011b; Martins et al., 2015), which has been implemented for modeling air vessels (Akpan et al., 2014).

2.5.3 Air pocket size

Air pockets, which directly affect filling and emptying processes, can accumulate in several parts in the system. In experimental facilities, different configurations regarding air pocket sizes have been conducted (Tijsseling et al., 2016; Coronado-Hernández et al., 2017b; Zhou et al., 2013a,b). However, the air pocket size cannot be controlled in actual installations. It is very important to note that small air pockets are more dangerous than large air pockets during these operations when air valves are not acting. Coronado-Hernández et al. (2018b) analyzed a 1200-m-long pipeline for filling and emptying procedures finding that the air pocket size is not sensitive over these events when air valves are acting. However, if there are no air valves along pipeline installations, the air pocket size can change remarkably the extreme values of absolute pressure. For instance, Izquierdo et al. (1999) shows a pipeline of a total length of 2000 m where a filling operation is analyzed. Results confirm that an air pocket size of 79 m produces a maximum absolute pressure head of 277 m, and considering an air pocket size of 140 m, the absolute pressure head is 191 m, which implies a difference of 30%.

2.5.4 Air valves behavior

Two sources of uncertainties were detected in the characterization of air valves produced by a lack of experimental tests of some manufacturers: (i) the first one corresponds to the air and vacuum flow rate curves, and (ii) the second one corresponds to the dynamic closure of air valves with entrapped air. Iglesias-Rey et al. (2014) analyzed discrepancies between manufacturer data and experimental tests considering different types of air valves.

On one hand, the development of experimental facilities to compute air and vacuum flow rate curves of air valves with large orifices is complex since an enough quantity of expelled/admitted air is required. Then, some manufacturers supply air and vacuum flow rate curves with discrepancies compared to experimental tests (Iglesias-Rey et al., 2014). An inappropriate characterization of air valves introduce an uncertainty in current models regarding inflow and outflow discharge coefficients.

On the other hand, the uncertainty associated to the dynamic closure with entrapped air in air valves is another deficiency of some manufacturers because they do not supply values of dynamic closure occurrence. The phenomenon occurs when outlet air velocity in air valves is too high (large volume of expelled air) producing a sustentation force on the float, which closes an air valve before the water arrives to this device and leaving a dangerous air pocket inside pipeline systems. The majority of air valves are prone to this phenomenon.

2.6 Practical considerations

2.6.1 General practices

In water pipelines, pressure surges and troughs of subatmospheric pressure can occur during filling and emptying processes, respectively, which are generated by entrapped air pockets in hydraulic systems. Air/vacuum valves (AVVs) should be installed along pipelines to prevent pressure surges, by exhausting air during filling processes and injecting a sufficient quantity of air during emptying processes, thus preventing subatmospheric conditions (AWWA, 2001; Ramezani et al., 2016). However, if AVVs are not correctly sized and well-maintained, problems can continue in installations (Ramezani et al., 2015; Stephenson, 1997; Fuertes, 2001; Tran, 2016).

Filling processes should be executed carefully, through slow maneuvers in the operation of discharge valves to produce an adequate expelled airflow rate. A pressure differential of 2 p.s.i (13,79 kPa) is recommended by the AWWA (2001) during this process. Water flow should be similar to air flow (Q_a) through AVVs with a water velocity of 0,3 m/s.

By contrast, a controlled emptying process should be performed replacing the water volume by an admitted air volume with a similar ratio to avoid dangerous troughs of absolute pressure (Coronado-Hernández et al., 2017b; Fuertes-Miquel et al., 2018b). (AWWA, 2001) recommends both water velocities from 0,3 to 0,6 m/s and a pressure differential of 5 p.s.i (34,5 kPa). If air valves are not installed or are undersized, then dangerous troughs of subatmospheric pressure occur, and hydraulic systems cannot be drained.

2.6.2 Air valves selection

The air valve selection during the air expulsion phase considers different aspects (Ramezani et al., 2015; AWWA, 2001). If air valves have been undersized, then the quantity of expelled air is not enough and peaks of pressure surge can occur. In contrast, for the larger air valve size, higher amounts of expelled air and water velocity are reached. If air valves have been oversized, then greater values of absolute pressure are reached compared to the situation without these devices. When the water reaches the air valve position, the air valve closes rapidly, producing a water hammer in a single-phase (water). Another consideration is the dynamic closure of air valves, which is typically not provided by manufacturers. When an air valve closes without expelling the entrapped air pocket completely, the water column compresses it, producing a dangerous pressure surge.

By contrast, during the emptying process, the air valve should be selected to protect the system from the troughs of subatmospheric pressure. In this sense, a larger air valve size reduces the lowest values of subatmospheric pressure.

2.6.3 Typical pipe selection practices

The pressure surges and troughs of subatmospheric pressure can be caused by a change in the water velocity during the filling and the emptying process, respectively (Wang et al.,

2017; Coronado-Hernández et al., 2017b). To select the pressure and the stiffness class in
pipelines, the water hammer effects should be computed not only for these operations but
also for hydraulic events such as the stopping of pumps (power failure), the rapid opening or
closing of valves, and pipe failure, among others. Extreme values of absolute pressure should
be selected to design the pipe characteristics.

The highest peak of the pressure surge reached in the aforementioned hydraulic events
should be used to select the pressure class in pipelines. Pipe manufacturers usually specify
the pressure class of water distribution networks depending on pipe material with typical
values varying from 6 to 16 bar (Mays, 1999). In other installations pressure class can be
different.

The stiffness class is selected according to the following conditions: (i) burial conditions
given by native soil, type of backfill, and cover depth, and (ii) the troughs of subatmospheric
pressure in the system. In this sense, designers can select the stiffness class based on manufacturer data, for instance, SN2500, SN5000, or SN10000.

2.7 Future research

This section provides future developments that should be considered to continue working in
this field, which are mentioned as follows:

1. Computational Fluid Dynamics (CFD) can be used to understand better the behavior of
   air pockets considering, among other things, the following circumstances: (i) the entrance of backflow air by drain valves during the emptying process, (ii) complex shapes
   of air-water interface, (iii) the variation of entrapped air pockets during these processes,
   (iv) the variation of thermodynamic proprieties of air inside air valves (García-Todolí
   et al., 2018), (v) the determination of absolute pressure caused by oversized air valves,
   and (vi) how an entrapped air causes a reduced cross section in pipelines.

2. The analysis of filling and emptying operations of water distribution networks has not
   been conducted neither numerically nor experimentally. The majority of experimental
   facilities correspond to single pipes or pipelines of undulating profiles. Also, the behavior of air in pipelines bifurcations needs to be analyzed since there are no studies
   related to the quantification of air volume fraction flowing to the downstream two pipe
   branches.

3. Current works consider the positing of air valves in high points of water pipelines;
   however, AWWA (2001) considers others locations of air valves which needs to be
   studied in points such as horizontal pipe branches, at vertical pumps, long descents and
   ascents, among others.

4. Many studies show that the determination of air and vacuum flow rate curves supplied
   by some manufactures is inadequate, as well as they do not supply reference values of
   dynamic closure occurrence. The correct air valves characterization is very important
   to analyze filling and emptying processes.

5. Current models of filling and emptying operations consider that polytropic model can be applied. However, there are some limitations of the polytropic model as reported by Graze et al. (1996).

2.8   Conclusions

This research presents current knowledge regarding the transient phenomena related to filling and emptying processes in water pipelines. A literature review was conducted on various mathematical models for the water phase, air-water interface, and air phase. Advantages and disadvantages of water phase models were described, as well as an identification of actual researchers. Air-water interface criteria are described, and the air phase and air valve characterization are explained. Four sources of uncertainty that need to be studied further, including the necessity to include a variable value for the polytropic coefficient and the friction factor, as well as the air pockets sizes and air valves selection in the system, were identified in current models. Typical practices about the air valve selection and the pressure and stiffness class in pipelines are summarized showing how emptying and filling operations should be considered for selecting them in hydraulic systems. Finally, future research on filling and emptying processes in pressurized pipelines have also been presented in this work.

Notation

a	= wave speed (m/s)
A	= cross-sectional area of pipe (m^2)
AV	= air valve (–)
A_{adm}	= cross sectional area of the air valve when air is admitted (m^2)
A_{exp}	= cross sectional area of the air valve when air is expelled (m^2)
C_{adm}	= inflow discharge coefficient (–)
C_{exp}	= outflow discharge coefficient (–)
c	= speed of sound (m/s)
D	= internal pipe diameter (m)
D_{av}	= air valve size (mm)
E	= internal energy (J)
f	= friction factor (–)
g	= gravity acceleration (m/s^2)
H	= piezometric head (m)
H_u	= upstream piezometric head (m)
H_d	= downstream piezometric head (m)
K	= polytropic coefficient in adiabatic conditions for air with a value of $K = 1{,}4$ (–)
K_d	= resistance coefficient of drain valve (m/(m^{3}s^{-1})^2)
K_u	= resistance coefficient of discharge valve (m/(m^{3}s^{-1})^2)
L	= length of the water column (m)
L_t	= total length of the pipe (m)
n	= polytropic coefficient (–)
m_a	= air mass (kg)
p_a^*	= absolute pressure of the air pocket (Pa)
p_{atm}^*	= atmospheric pressure (Pa)
t	= time (s)
T	= temperature ($^\circ K$)
R	= constant air (287 J / kg / $^\circ$K)
Q_e	= net quantity of heat (J)
Q_a	= air flow (m^3/s)
Q_w	= water discharge (m^3/s)
X	= distance along of a pipe (m)
x_0	= initial air pocket size (m)
V_a	= air volume (m^3)
v_w	= water velocity (m/s)
v_a	= air velocity (m/s)
W	= work done by the system (J)
ρ_a	= air density (kg/m^3)
θ	= pipe inclination (rad)
Δz	= difference elevation (m)
0	= refers to an initial condition

Chapter 3

Transient phenomena during the emptying process of a single pipe with water-air interaction

This chapter is extracted from the paper:

Transient phenomena during the emptying process of a single pipe with water-air interaction
Coauthors: Fuertes-Miquel, V. S; Coronado-Hernández O. E; Iglesias-Rey P. L.; Mora-Meliá, D.
Journal: Journal of Hydraulic Research ISSN 0022-1686
2017 Impact Factor: 2.076. Position JCR 41/128 (Q2). Civil Engineering
State: Published. Latest Articles. 2018; DOI: 10.1080/00221686.2018.1492465.

3.2   Introduction

Over recent decades, the analysis of transient phenomena for a single-phase liquid has been
studied extensively by researchers and there are currently numerous models that represent
correctly the water behavior in any hydraulic system. However, transient flow with trapped
air is much more complex to simulate because there are two fluids (water and air) in two different phases (liquid and gas) (Fuertes, 2001), which implies a higher complexity to establish
reliable mathematical models.

Filling and emptying maneuvers are common pipelines operations with trapped air. These
operations are repeated periodically and must be considered to avoid future problems associated to extreme values of air pocket pressure. For this reason, many researchers have focused
their efforts on the development of different models for the analysis on entrapped air in water supply networks (Zhou et al., 2013a,b; Wang et al., 2016; Martins et al., 2015; Laanearu
et al., 2012).

To date, researchers have mostly focused on filling processes, developing models that
have shown great accuracy in comparison to experimental data. In this sense, early studies on
pipelines were conducted by Liou and Hunt (1996) and Izquierdo et al. (1999), and Zhou et al.
(2013a,b) developed recently models to analyze the process of rapid filling. Additionally, a
few pieces of research have been published analyzing the influence of different parameters
through experimental investigations of the systems (Hou et al., 2014; Zhou and Liu, 2013).Fi-
nally, others studies have focused on techniques and methods for solving the filling problem
(Wang et al., 2016; Martins et al., 2015; Malekpour and Karney, 2014; Liu et al., 2011; Martins et al., 2010b), analyzing the consequences of the formation of air pockets (Pozos et al.,
2010), and the occurrence of cavitation in the propagation of transient phenomena (Wang
et al., 2003; Guinot, 2001).

Emptying processes may also be critical, but they have received less attention from researchers. On this matter, Laanearu et al. (2012) developed a semi-empirical model for es-
timating local energy losses and Tijsseling et al. (2016) presented a model for the emptying
process with pressurized air in a pipeline. Nevertheless, there is a lack of studies on the water
draining maneuvers in engineering applications of real hydraulic systems.

Therefore, this paper presents a new mathematical model to be used for emptying processes. The proposed model simulates the liquid column with a rigid model using the mass
oscillation equation (Cabrera et al., 1992; Liou and Hunt, 1996; Izquierdo et al., 1999; Lee,
2005), which provides sufficient accuracy by using a moving air-water interface (Izquierdo
et al., 1999; Zhou et al., 2013a). Additionally, it is possible to consider thermodynamic behaviors such as: (a) an isothermal process, where the temperature remains constant inside the
hydraulic installation; (b) an adiabatic process, where the hydraulic installation exchanges
no heat with its surroundings; and (c) an intermediate processes, which involves temperature
change and heat transfer. In this sense, the relationship between the absolute pressure and total volume of entrapped air characterizes the behavior of air, which is known as a polytropic
process. If the transient process occurs slowly enough, it is considered isothermal, and the
polytrophic coefficient is 1,0. In contrast, if the transient process is fast enough, it is consid-

ered adiabatic, and the polytrophic coefficient is 1,4 (Martins et al., 2015; Zhou et al., 2013b; Martin, 1976; Izquierdo et al., 1999; Fuertes-Miquel et al., 2016). In actual installations, isothermal or adiabatic processes are rare, so an intermediate situation occurs.

This work applies the developed model to the emptying process in two cases. On one hand, Case No. 1 corresponds to a single pipe with the upstream end closed without an air valve. As such, it is not possible to empty the pipeline, so subatmospheric pressure troughs are generated and may collapse the system. In this case, a polytrophic evolution of the air pocket is used (Zhou et al., 2013b). On the other hand, Case No. 2 corresponds to a single pipe with an air valve installed on the upstream end, which protects the system during the emptying process. As there is air inlet, this case also needs the continuity equation of the air pocket, the expansion-compression equation of the entrapped air pocket (Leon et al., 2010; Martin, 1976), and the air valve characterization (Iglesias-Rey et al., 2014; Wylie and Streeter, 1993). Water column separation cannot be predicted by this model, but such a situation rarely occurs in normal water emptying processes. In each case, systems of differential-algebraic equations (DAE) are solved using numerical methods. Computational simulations were performed using the Simulink Library in Matlab. The proposed model was validated by comparing the computed pressure oscillation patterns with measured results. This study shows how the model fits the experimental data, and the results can be used to develop practical applications for emptying processes.

3.3   Mathematical model

This section presents a mathematical model for analyzing the emptying process in a single pipe. Figure 3.1 shows the configuration of a single pipe for the two cases analyzed. Case No. 1 does not have an air valve installed, while in Case No. 2, an air valve is used to increase reliability during the emptying process.

3.3.1   Case No. 1: A simple pipe with the upstream end closed

The analysis of the emptying process in a simple pipe with the upstream end closed presents an air pocket inside the installation (see Fig. 3.1a).

The following assumptions are made:

- Water behavior has been modelled using the rigid model approach.

- Slope, diameter and pipe roughness have been considered constant.

- A constant friction factor is considered to account for losses using the Darcy-Weisbach equation (Liou and Hunt, 1996; Izquierdo et al., 1999; Zhou et al., 2013a; Laanearu et al., 2012; Tijsseling et al., 2016).

- A polytrophic coefficient is used to model the behavior of entrapped air.

- A valve is installed at the downstream end.

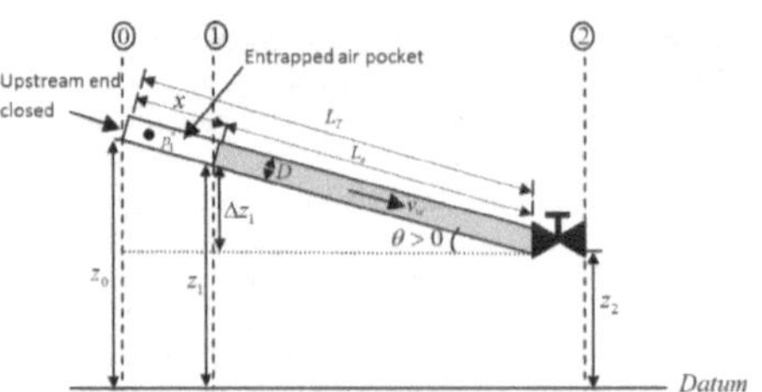

(a) Case No. 1: A simple pipe with upstream end closed

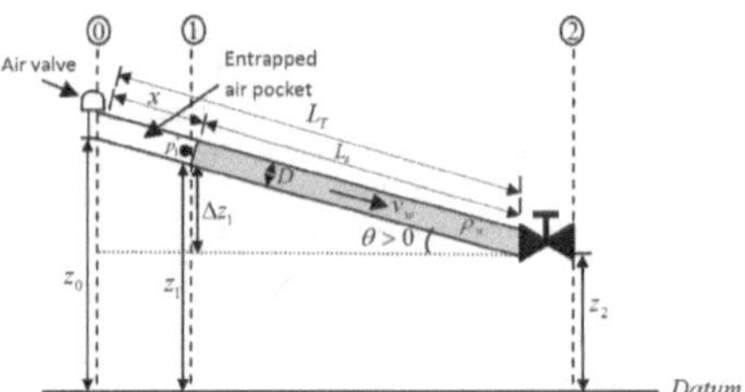

(b) Case No. 2: A simple pipe with an air valve installed in the upstream end

Figure 3.1: Schematic of an entrapped air pocket in a single pipe while water empties

- The air-water interface is a well-defined cross section. This assumption depends on the pipe diameter, the water velocity and the pipe slope.

- The pipe can resist dangerous troughs of subatmospheric pressure during the transient phenomenon.

Under these hypotheses, the problem is modelled by:

Emptying column

- Mass oscillation equation for emptying column (rigid column approach) (Fuertes, 2001):

$$\frac{dv_w}{dt} = \frac{p_1^* - p_{atm}^*}{\rho_w L_e} + g\frac{\Delta z_1}{L_e} - f\frac{v_w|v_w|}{2D} - \frac{R_v g A^2 v_w|v_w|}{L_e} \tag{3.1}$$

where v_w = water velocity, p_1^* = absolute pressure of air pocket, p_{atm}^* = atmospheric pressure, ρ_w = water density, L_e = length of emptying column, Δz_1 = elevation difference, g = gravity acceleration, f = Darcy-Weisbach friction factor, D = internal pipe diameter, A = cross sectional area of pipe, R_v = resistance coefficient, and Q_w = water discharge.Minor losses in the valve were estimated by using the formulation $h_m = R_v Q_w^2$.

- Interface position of the emptying column:

$$\frac{dL_e}{dt} = -v_w \ \left(L_e = L_{e,0} - \int_0^t v_w \mathrm{d}t \right) \tag{3.2}$$

where $L_{e,0}$ = initial value of L_e.

Air pocket

- Entrapped air pocket:

$$p_1^* V_a^k = p_{1,0}^* V_{a,0}^k \quad or \quad p_1^* x^k = p_{1,0}^* x_0^k \tag{3.3}$$

where V_a = air volume, $V_{a,0}$ = initial air volume, $p_{1,0}^*$ = initial value of p_1^*, k = polytropic coefficient, x = air pocket size, and x_0 = initial length of x.

In summary, a 3x3 system of DAE (equations 3.1 - 3.3) describes the whole problem. The DAE, with the corresponding boundary and initial conditions, can be solved for the 3 unknowns (v_w, L_e and p_1^*).

Initial and boundary conditions

With the system being at rest ($t = 0$), the initial length of the emptying column and the size of the air pocket defined, the initial conditions are described by $v_w(0) = 0$, $L_{e,0} = L_T - x_0$ and $p_{1,0}^* = p_{atm}^* = 101325$ Pa.

The upstream boundary condition is given by $p_{1,0}^*$ (initial condition of air pocket), and the downstream boundary condition is given by p_{atm}^* (water free discharge to the atmosphere).

Gravity term

The gravity term ($\Delta z_1 / L_e$) in equation 3.1 is always constant and can be modeled by the following equation:

$$\frac{\Delta z_1}{L_e} = \sin(\theta) \tag{3.4}$$

3.3.2 Case No. 2: A simple pipe with an air valve installed in the up-stream end

The analyses of the emptying process in a simple pipe with an air valve installed at the
upstream end is shown in Fig. 3.1b. The assumptions are the same as in Case No. 1. Consequently, equations 3.1 and 3.2 were used to represent the behavior of emptying the column.
The air pocket was modelled using the following equations:

- Continuity equation: This equation was used, considering that the air density of air
  pocket (ρ_a) and the air density in air valve ($\rho_{a,nc}$) are different (see Fig. 3.2).

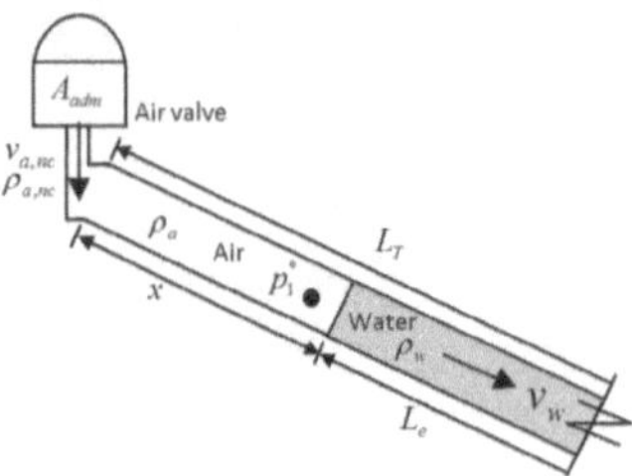

Figure 3.2: Location of air valve

Applying the continuity equation:

$$\frac{dm_a}{dt} = \rho_{a,nc} v_{a,nc} A_{adm}$$ (3.5)

and by deriving the expression:

$$\frac{dm_a}{dt} = \frac{d(\rho_a V_a)}{dt} = \frac{d\rho_a}{dt} V_a + \frac{dV_a}{dt} \rho_a$$ (3.6)

where m_a = air mass, $v_{a,nc}$ = air velocity in normal conditions admitted by the air
valve, and A_{adm} = cross sectional area of the air valve.

If $V_a = xA = (L_T - L_e)A$ and $dV_a/dt = -(dL_e/dt)A = vA$, then:

$$\frac{d\rho_a}{dt} = \frac{\rho_{a,nc} v_{a,nc} A_{adm} - v_w A \rho_a}{A(L_T - L_e)}$$ (3.7)

- Expansion-compression equation (Leon et al., 2010; Martin, 1976; Lee, 2005):

$$\frac{dp_1^*}{dt} = -k\frac{p_1^*}{V_a}\frac{dV_a}{dt} + \frac{p_1^*}{V_a}\frac{k}{\rho_a}\frac{dm_a}{dt}$$ (3.8)

- Air valve characterization (Iglesias-Rey et al., 2014; Wylie and Streeter, 1993): The mass of air flowing through the air valve depends on the value of the atmospheric pressure (p^*_{atm}) and the absolute pressure of air pocket (p^*_1).

  – Subsonic air flow in ($p^*_{atm} > p^*_1 > 0,528p^*_{atm}$):

  $$Q_{a,nc} = C_{d,adm}A_{adm}\sqrt{7p^*_{atm}\rho_{a,nc}\left[\left(\frac{p^*_1}{p^*_{atm}}\right)^{1,4286} - \left(\frac{p^*_1}{p^*_{atm}}\right)^{1,714}\right]} \qquad (3.9)$$

  – Critical air flow in ($p^*_1 \leq 0,528p^*_{atm}$):

  $$Q_{a,nc} = C_{d,adm}A_{adm}\frac{0,686}{\sqrt{RT}}p^*_{atm} \qquad (3.10)$$

  where $Q_{a,nc}$ = air flow in normal conditions across the air valve, $C_{d,adm}$ = inflow discharge coefficient, R = air constant, and T = air temperature.

In summary, a 5x5 system of DAE (equations 3.1, 3.2, 3.7, 3.8 and 3.9 or 3.10) describe the whole problem. Together with the corresponding boundary and initial conditions, 5 unknowns can be solved (v_w, L_e, p^*_1, $v_{a,nc}$ and ρ_a).

Initial and boundary conditions

The initial conditions are described by $v_w(0) = 0$, $L_{e,0} = L_T - x_0$, $p^*_{1,0} = p^*_{atm} = 101325$ Pa, $\rho_{a,0} = 1,205$ kg/m^3 and $v_{a,nc}(0) = 0$. The boundary conditions are the same as in Case No. 1.

3.4 Model verification

The proposed model was applied to a methacrylate pipeline that was 42 mm in internal diameter and 4,36 m long. Experiments were carried out at the Fluids Laboratory located at the Universitat Politècnica de València (Valencia, Spain). The installation setup consisted of a water supply tower, a methacrylate pipe of 4,16 m, a PVC pipe of 0,2 m, PVC joints, two ball valves with a 42 mm internal diameter located at the ends and a free-surface basement reservoir used to collect the drainage water. Initially, the methacrylate pipeline was filled, and the emptying process was started by the opening of a ball valve. A transducer was installed upstream to measure the pressure. Figure 3.3 shows the experimental setup.

The gravity term for this schematic configuration (see Fig. 3.3a) depends on the position of the emptying column. There are two possibilities:

1. When the air-water front in the emptying column has not reached the PVC pipe ($L_e >= L_p$):

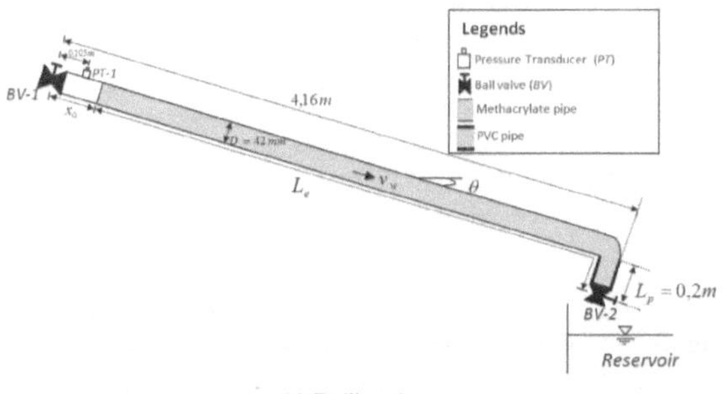

(a) Facility scheme

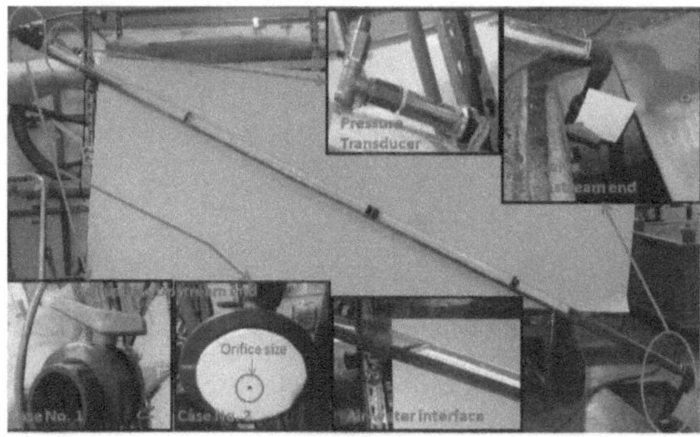

(b) Details of experimental setup

Figure 3.3: Schematic of experimental setup

Table 3.1: Case No. 1 - Experimental setup - Characteristics of tests 1 to 12

Test No.	x_0 (m)	θ (rad)	R_v x 10^{-6}(ms^{2}m^{-6})	T_m (s)
1	0,205	0,457	11,89	0,40
2	0,340	0,457	11,89	0,40
3	0,450	0,457	11,89	0,50
4	0,205	0,457	25,00	0,25
5	0,340	0,457	22,68	0,15
6	0,450	0,457	30,86	0,30
7	0,205	0,515	14,79	0,50
8	0,340	0,515	14,79	0,40
9	0,450	0,515	14,79	0,75
10	0,205	0,515	135,21	0,30
11	0,340	0,515	138,41	0,30
12	0,450	0,515	100,00	0,30

$$\frac{\Delta z_1}{L_e} = \frac{L_e - L_p}{L_e} \sin(\theta) + \frac{L_p}{L_e} \cos(\theta) \tag{3.11}$$

2. When the air-water front in the emptying column has reached the PVC pipe ($L_e < L_p$):

$$\frac{\Delta z_1}{L_e} = \cos(\theta) \tag{3.12}$$

To verify the proposed model, comparisons of the computed and measured pressure oscillation patterns and trough pressures were conducted. A friction factor $f = 0,018$ was used. During the experiments, various air pocket sizes (x_0), pipe slopes (θ), resistance coefficients (R_v) and valve maneuvering times (T_m) were tested. A synthetic maneuver with a resistance coefficient was modeled as the opening of the ball valve.

The results were obtained in both cases using an adiabatic process ($k = 1,4$), which means that the experiments presented a fast transient phenomenon. For the Case No. 1, twelve tests were performed (see Table 3.1). For the Case No. 2, twenty four tests were performed (twelve with $D_{av} = 1,5$ mm and twelve with $D_{av} = 3,0$ mm). Each test was repeated twice to confirm the measurement.

For Case No. 1, the ball valve located upstream was closed before the start of the simulation. Table 3.1 displays tests 1 through 12 for Case No. 1.

Figures 3.4 and 3.5 show the oscillations during the first three seconds of all tests. The proposed model can predict the pressure oscillation patterns and the associated trough subatmospheric pressures. For instance, test 2 is characterized by a few waves during the first three seconds, which are produced by the opening of the ball valve ($R_v = 11,89 \bullet 10^6$ ms^{2}m^{-6}), as shown in Fig. 3.4b. Additionally, the influence of the valve maneuvering time can be seen by the lag in the pressure oscillation pattern between Repetition 1 and Repetition 2.

The minimum subatmospheric pressure head of 8,23 m was reached. In contrast, test 10 has practically no waves during the transient (see Fig. 3.4j) and a high flow resistance coefficient value ($R_v = 135,21 \bullet 10^6$ ms^{2}m^{-6}) where the trough of the subatmospheric pressure head of 8,14 m was found.

In Case No. 2, an orifice was made upstream to represent the air valve (see Figures 3.6, 3.7, 3.8 and 3.9). Two different orifice sizes were used to show the impact of the air inlet. The smallest orifice size corresponds to the lowest value of the inflow discharge coefficient. The inflow discharge coefficients for $D_{av} = 1,5$ mm and $D_{av} = 3,0$ mm are 0,55 and 0,65, respectively. These values have been calibrated. Figure 3.10 shows the experimental results and the characteristic curves for the two orifice sizes. The formulation presented by Wylie and Streeter (1993) predicts the experimental data. Table 3.2 shows the tests.

Figures 3.6 and 3.8 show a comparison between the computed and measured absolute pressure oscillation patterns for tests 13 to 36. Again, the proposed model shows a fairly good overall agreement with the experiments. The subatmospheric pressure trough occurs immediately when the ball valve is opened. Then absolute pressure starts to increase slowly until it reaches atmospheric conditions. For instance, the trough of the subatmospheric pressure head for test 14 (8,89 m) is lower than test 32 (9,51 m). Test 14 displays a transient longer than test 32.

Finally, Fig. 3.12 shows a comparison between the calculated and measured subatmospheric pressure troughs for both cases. For all tests, the proposed model predicts the trough of the subatmospheric pressure, which is very important to ensure pipeline safety. The main risk is presented in Case No. 1 when there no air inlet. Case No. 2 shows that an orifice of size $D_{av} = 1,5$ mm reaches a greater subatmospheric pressure trough than an orifice of size $D_{av} = 3,0$ mm.

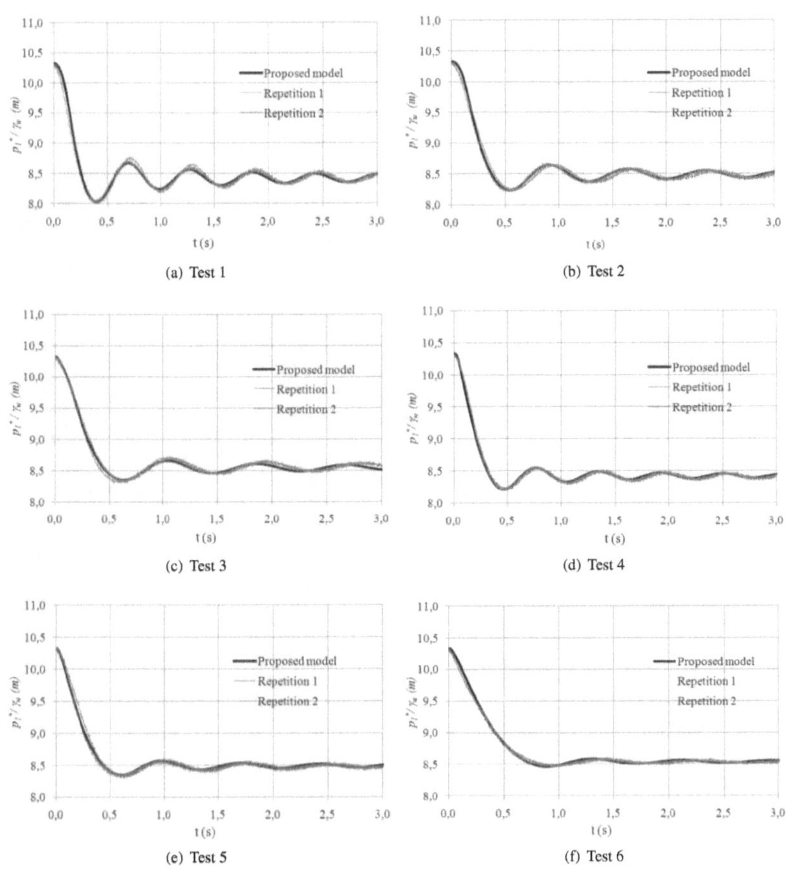

Figure 3.4: Case No. 1 - Comparisons between calculated and measured absolute pressure of the air pocket for tests 1 to 6

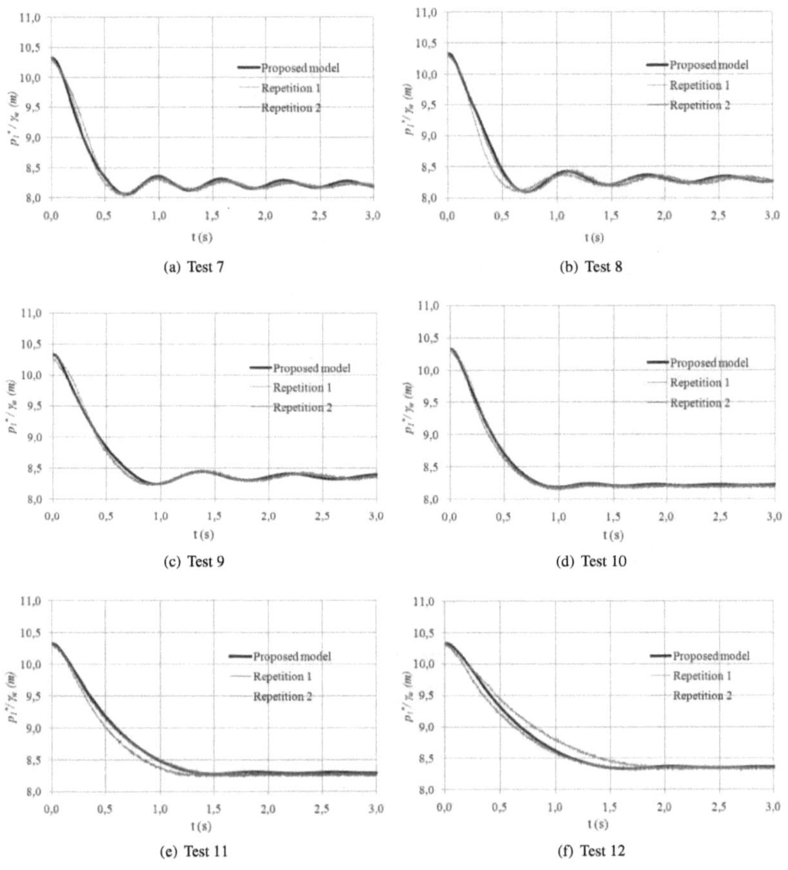

(a) Test 7

(b) Test 8

(c) Test 9

(d) Test 10

(e) Test 11

(f) Test 12

Figure 3.5: Case No. 1 - Comparisons between calculated and measured absolute pressure
of the air pocket for tests 7 to 12

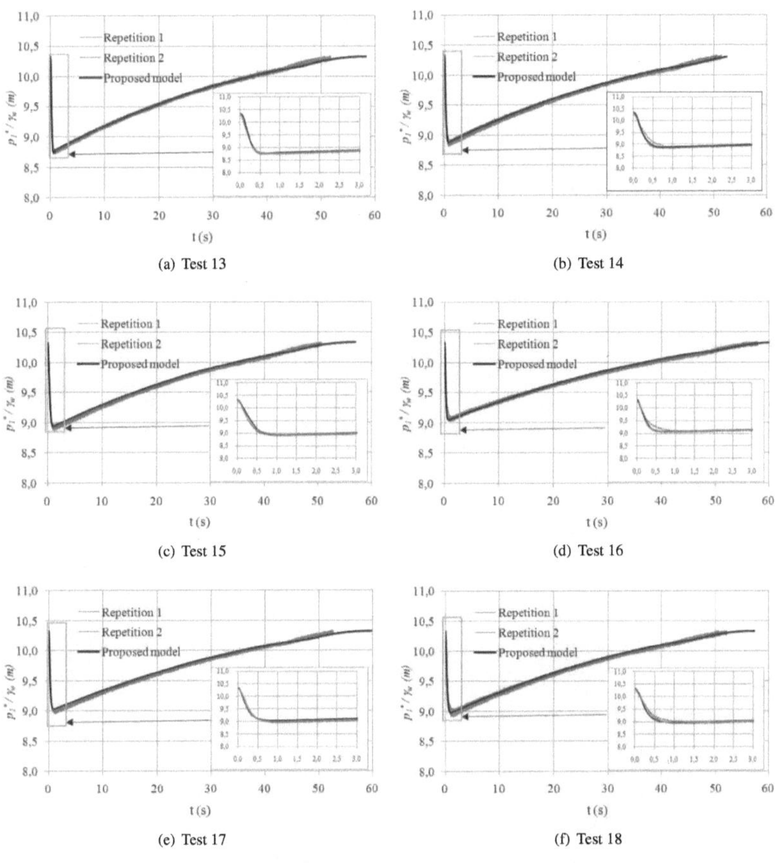

(a) Test 13

(b) Test 14

(c) Test 15

(d) Test 16

(e) Test 17

(f) Test 18

Figure 3.6: Case No. 2 - Comparisons between calculated and measured absolute pressure of the air pocket for tests 13 to 18 ($D_{av} = 1{,}5$ mm)

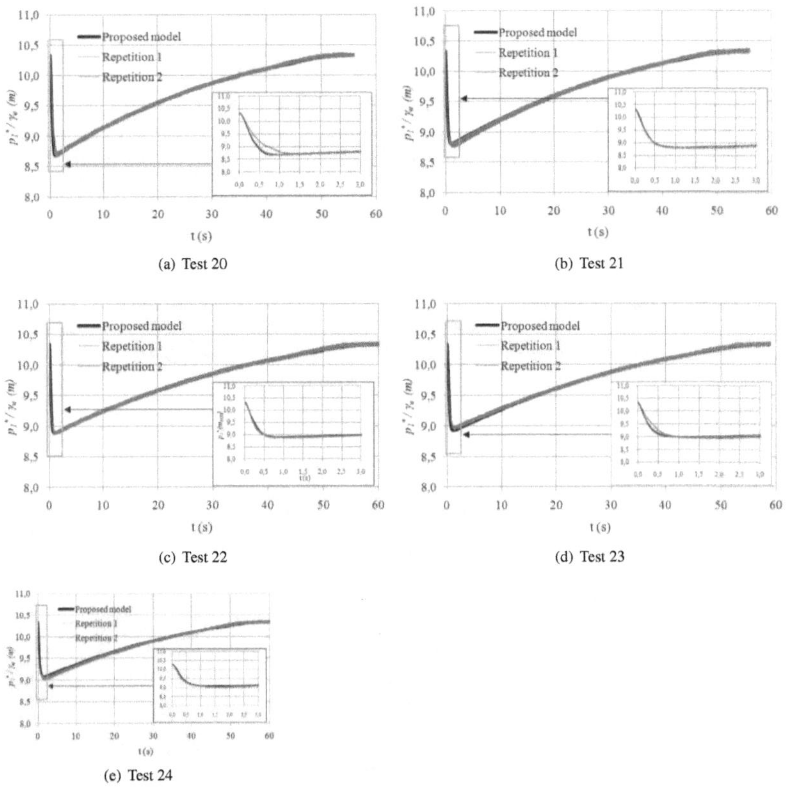

(a) Test 20

(b) Test 21

(c) Test 22

(d) Test 23

(e) Test 24

Figure 3.7: Case No. 2 - Comparisons between calculated and measured absolute pressure
of the air pocket for tests 19 to 24 ($D_{av} = 1,5$ mm)

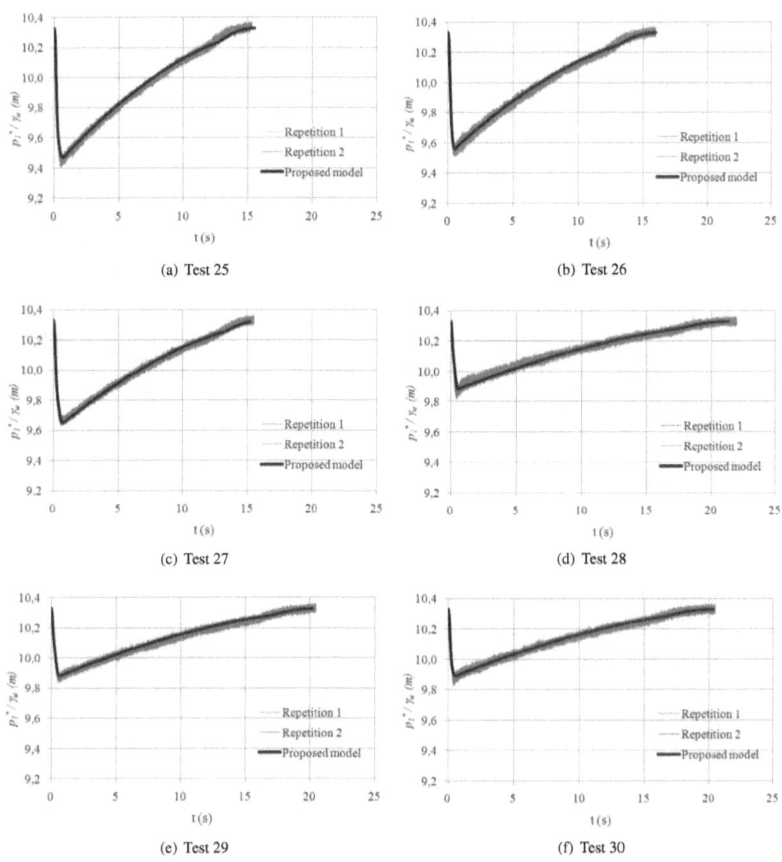

(a) Test 25 (b) Test 26

(c) Test 27 (d) Test 28

(e) Test 29 (f) Test 30

Figure 3.8: Case No. 1 - Comparisons between calculated and measured absolute pressure of the air pocket for tests 25 to 30 ($D_{av} = 3,0$ mm)

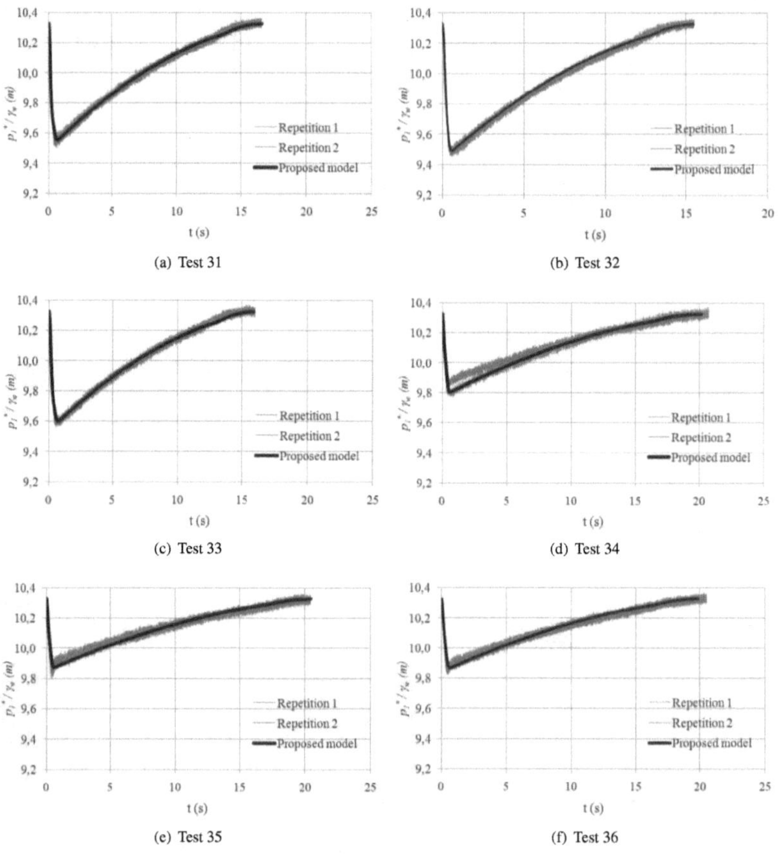

Figure 3.9: Case No. 1 - Comparisons between calculated and measured absolute pressure
of the air pocket for tests 31 to 36 ($D_{av} = 3{,}0$ mm)

Table 3.2: Case No. 2 - Experimental setup - Characteristics of tests 13 to 36

Test No.	D_{av} (mm)	x_0 (m)	θ (rad)	$R_v \bullet 10^{-6}$ (ms^{2}m^{-6})	T_m (s)
13	1,5	0,205	0,457	9,18	0,40
14	1,5	0,340	0,457	11,11	0,40
15	1,5	0,450	0,457	11,11	0,40
16	1,5	0,205	0,457	20,66	0,25
17	1,5	0,340	0,457	16,66	0,15
18	1,5	0,450	0,457	12,76	0,10
19	1,5	0,205	0,515	8,65	0,40
20	1,5	0,340	0,515	9,18	0,50
21	1,5	0,450	0,515	11,11	0,40
22	1,5	0,205	0,515	18,90	0,35
23	1,5	0,340	0,515	17,36	0,30
24	1,5	0,450	0,515	22,68	0,25
25	3,0	0,205	0,457	2,16	0,50
26	3,0	0,340	0,457	2,60	0,50
27	3,0	0,450	0,457	2,97	0,50
28	3,0	0,205	0,457	6,25	0,40
29	3,0	0,340	0,457	5,95	0,40
30	3,0	0,450	0,457	5,95	0,35
31	3,0	0,205	0,515	3,08	0,50
32	3,0	0,340	0,515	2,60	0,35
33	3,0	0,450	0,515	3,08	0,50
34	3,0	0,205	0,515	5,67	0,40
35	3,0	0,340	0,515	6,57	0,40
36	3,0	0,450	0,515	6,25	0,40

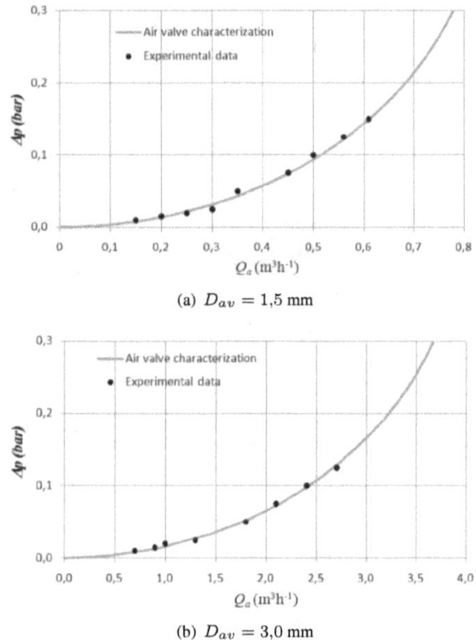

(a) $D_{av} = 1,5$ mm

(b) $D_{av} = 3,0$ mm

Figure 3.10: Laboratory test results of air valves during
air admission

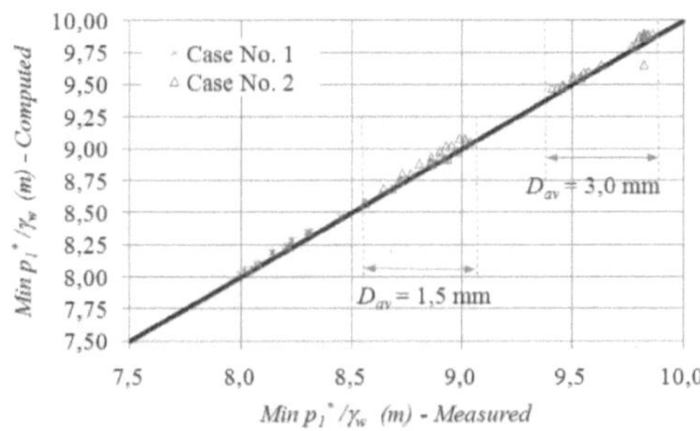

Figure 3.11: The comparison between computed and measured minimum pressures

3.5 Case study and results

To show what has been set forth, and to quantify the order of magnitude of some of the terms used, we present practical applications for the two cases mentioned. For both cases, a polytrophic coefficient $k = 1,2$ has been selected, which is a typical value in practical applications.

3.5.1 Case No. 1: A simple pipe with the upstream end closed

The system presented in Fig. 3.1a has been solved with the following data: $L_T = 600m$, $f = 0,018$, $D = 0,35$ m; $\Delta z_1 = 15$ m, $R_v = 0,06$ ms^{2}m^{-6}, $x_0 = 200$ m, $k = 1,2$, and $p_{1,0}^* = 101325$ Pa. The results obtained are shown in Fig. 3.12.

According to the results, the authors can deduce:

- Fig. 3.12a shows that the water flow increases rapidly until reaching a maximum value of 0,26 m^{3}s^{-1} at 23,5 s. Subsequently, the flow of water decreases until it reaches a value of 0 m^{3}s^{-1} at 123,6 s. Then it continues to decrease until $-0,06$ m^{3}s^{-1} at 158,8 s, indicating negative velocities occur. Hence, a setback occurs while emptying the column. Finally, oscillations in the water flow are observed until it reaches a value of 0 m^{3}s^{-1}, when draining is stopped, and part of the water remains in the installation.

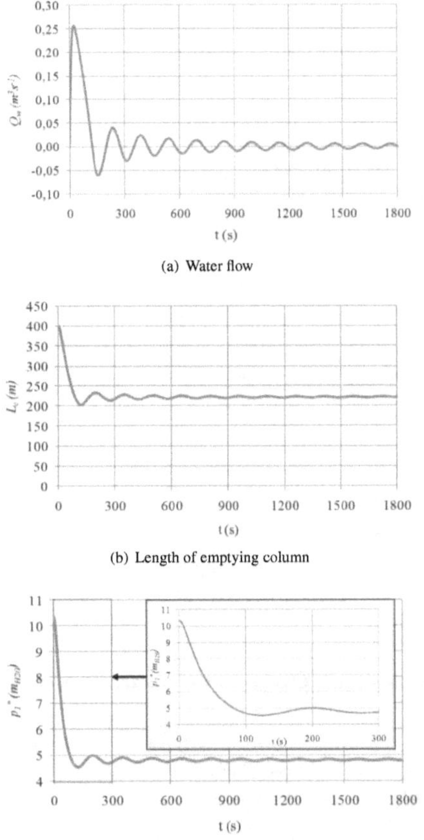

(a) Water flow

(b) Length of emptying column

(c) Absolute pressure of air pocket

Figure 3.12: Results of Case No. 1 - Full transient analysis
of the system under consideration

- The water in the column decreases rapidly at 123,6 s at the beginning of the transient phenomenon because the water has been draining quickly through the valve. Subsequently, small oscillations occur along the length of the emptying column. The final length of the column is 221 m, which means that only 179 m could be drained (see Fig. 3.12b).

- Initially, the air in the pocket is at atmospheric pressure (101325 Pa). When the valve is opened, the air pressure head decreases rapidly in the pocket until it reaches a value of 4,8 m. This value continues indefinitely (see Fig. 3.12c).

It is also important to note that the proposed model does not simulate the backflow air phenomemon which is explained in next chapter.

3.5.2 Case No. 2: A simple pipe with an air valve installed in the upstream end

The system is presented in Fig. 3.1b. The data are the same as in Case No. 1. In addition, $D_{av} = 50$ mm ($A_{adm} = 0,0019$ m^2), and $C_{d,adm} = 0,50$. The results obtained are shown in Fig. 3.13.

From the results, it can be deduced that the maximum water flow of 0,27 m^{3}s^{-1} in the transient event occurs at 29,5 s, as shown in Fig. 3.13a. At the beginning of the transient event, there is less air flow than water flow because the valve restricts the air flow rate. At 125,2 s, the air flow surpasses the water flow because a large part of the column length has been drained, which corresponds to a decrease in the output water flow. From 125,2 to 291,2 s, water flow oscillations occur because the water flow rate is now less than the air flow rate (see Fig. 3.13a). Figure 3.13b shows that the emptying column is drained completely in 291,2 s. During the first 125,2 s, the length of the emptying column decreases because the water flow is greater than the air flow admitted by the air valve. When oscillations occur in the water flow, the pipe drains more slowly.

The change in absolute pressure parallels the air density in the air pocket (see Fig. 3.13c). In the beginning, the pressure and density decrease until they reach minimum values of 8,19 m and 0,99 kgm^{-3}, respectively, which occur at 86,65 s. These variables increase until they reach atmospheric conditions. Figure 3.13d presents the components of total energy per unit weight (or total head). The velocity head, potential head, and absolute pressure head are based on Fig. 3.13a, Fig. 3.13b, and Fig. 3.13c, respectively. The total head starts with a value of 20,33 m, where the water velocity is null, the air pocket is at atmospheric conditions (101325 Pa), and the initial potential head is 10,0 m. The velocity head has no relevance during the hydraulic event. The total head is the sum of the absolute pressure head and the potential head. The total head decreases from 0 to 150 s. From 150 to 291,2 s, the total head decreases slowly with values close to atmospheric conditions (101325 Pa or 10,33 m).

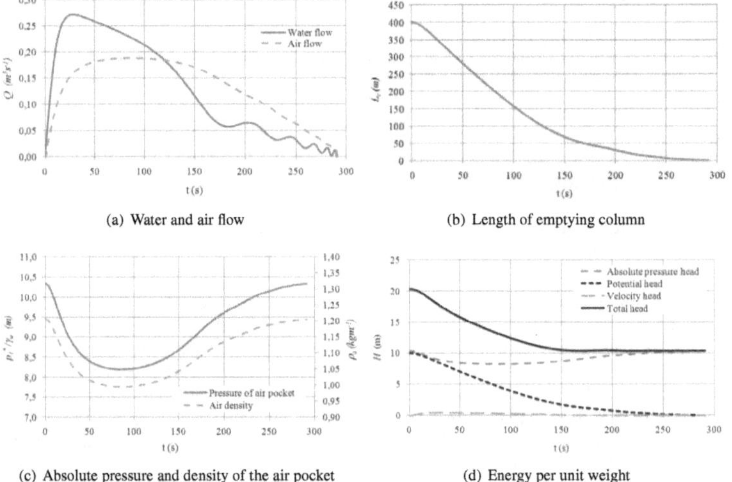

(a) Water and air flow

(b) Length of emptying column

(c) Absolute pressure and density of the air pocket

(d) Energy per unit weight

Figure 3.13: Results of Case No. 2 - Full transient analysis of the system under consideration

Table 3.3: Case No.1 - Influence of air pocket sizes

Le_0 (m)	x_0 (m)	Min p_1^* (m)	t (s)
550	50	1,39	167
500	100	2,59	148
450	150	3,62	142
400	200	4,53	121
350	250	5,37	109
300	300	6,15	98

Table 3.4: Influence of polytropic coefficient (k)

k	Min p_1^* (m)	Min Le (m)	Max Q_w (m^{3}s^{-1})
1,0	4,99	186	0,260
1,2	4,53	203	0,256
1,4	4,15	217	0,252

3.5.3 Sensitivity analysis

Case No.1: A simple pipe with upstream end closed

It is important to note the great influence of the air pocket size on the minimum of the subatmospheric pressure. Table 3.3 shows the results taking different air pocket sizes. The smaller the air pocket, the lower values of subatmospheric pressure are achieved. Values of the Subatmopheric pressure head have found in the range of 1,39 and 6,15 m .

It is also important to analyze the influence of the opening times of the drain valve which was simulated introducing a synthetic maneuver.

Figure 3.14 shows the results considering different opening times (T_m) in the range of 0 s and 300 s. The slower the opening times, the lower the maximum water flow (see Figure 3.14a). The opening times do not affect the troughs of the subatmospheric pressure, as shown in Figure 3.14b.

Table 3.4 presents the results of the influence of the polytropic coefficient (k). Typical values of this coefficient are 1,0, 1,2 and 1,4. The higher the polytropic coefficient, the lower values of the subatmospheric pressure but the higher length of emptying column. The differences on the water flow are negligible.

Case No.2: A simple pipe with an air valve installed in the upstream end

The influence of opening times (T_m) in the drain was analyzed considering a range of 0 s and 300 s, see Figure 3.15. The slower the opening times, the lower the water flow are (see Figure 3.15a). Troughs of the subatmospheric pressure head are found from 8,19 to 8,44 m (see Figure 3.15b).

It is worth underscoring here the great influence of the selecting the appropriate air valve

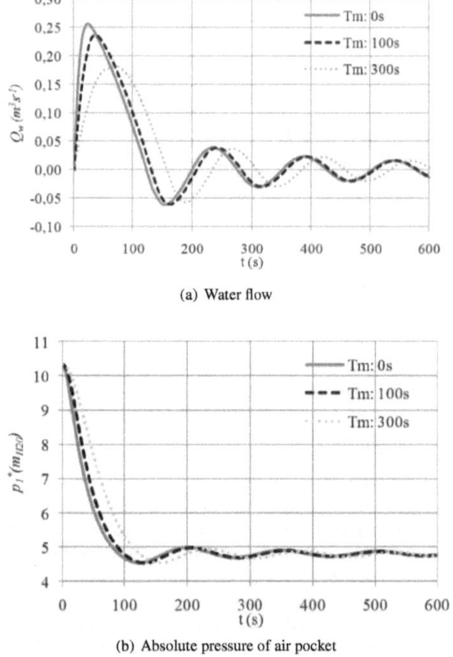

(a) Water flow

(b) Absolute pressure of air pocket

Figure 3.14: Case No.1 - Influence of the opening times in the drain valve

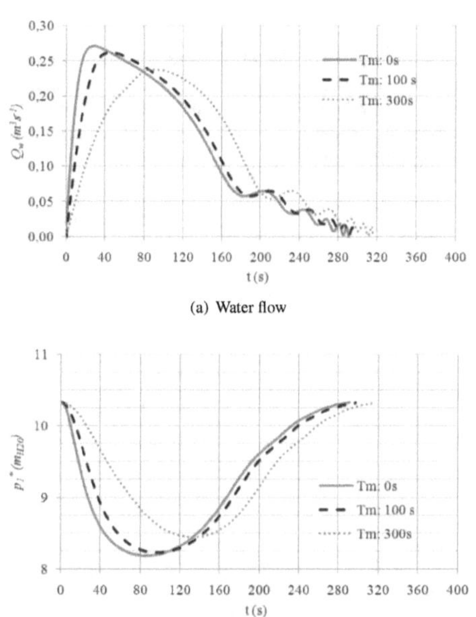

(a) Water flow

(b) Absolute pressure of air pocket

Figure 3.15: Case No.2 - Influence of the opening time in the valve

Table 3.5: Characteristics of the air valves

D_{av} (mm)	$C_{d,adm}$	A_{adm} (m^2)
25	0,50	0,00049
50	0,50	0,00196
100	0,50	0,00785

size. Table 3.5 summarizes the characteristics of the used air valves. Figure 3.16 shows the results for the water flow and the absolute pressure. According to results, engineers should select an air valve with $D_{av} = 100$ mm in order to reduce problems related to the subatmospheric pressure occurrence. An inappropriate selection of the air valve size, produces the lower values of the subatmospheric pressure.

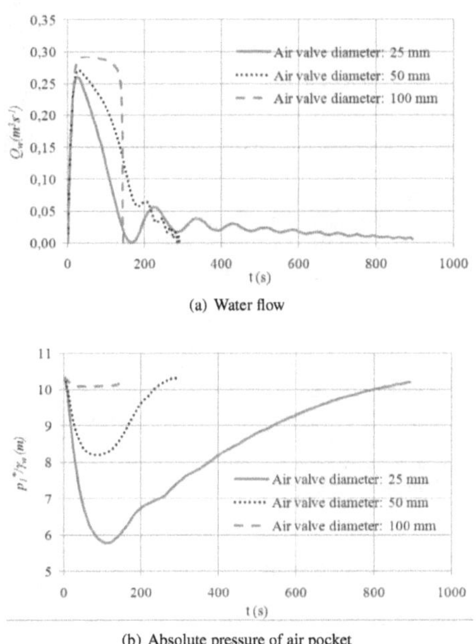

(a) Water flow

(b) Absolute pressure of air pocket

Figure 3.16: Case No.2 - Influence of the air valve diameter

3.6 Conclusions

Traditionally, emptying processes have received little attention from researchers. However, the development of models to simulate the transient behavior in these processes is considered to be of great interest to the scientific community. In this chapter, a mathematical model for analyzing the transient phenomenon during the emptying process of a single pipe is developed.

First, the proposed model was validated using an experimental facility. A comparison between computed and measured absolute pressure oscillation pattern shows that the proposed model accurately depicts this phenomenon. The proposed model can predict the subatmospheric pressure pattern, which is very important for pipe stiffness and air valve selection.

Finally, the proposed model was applied in a theoretical case to quantify the order of magnitude of the main hydraulic variables. Based on the analysis, the following conclusions can be drawn: (i)during the emptying process of a single pipe without air valves (Case No. 1) troughs of subatmospheric pressure are generated that could cause the collapse of the system. Additionally, the total length of the emptying column will not completely drain. To avoid these problems, air valves must be installed at the high points of the installation (Case No. 2). If the air valve size is suitable, i.e., it has been well designed, the pipe displays a small trough of subatmospheric pressure. However, if the air valve size is inappropriate, the amount of air required does not enter into the system and important troughs of subatmospheric pressure can be reached. As a consequence, the system could collapse; and (ii) it is recommended that great care is taken during the emptying process, especially when an entrapped air pocket is present in the system. Slow maneuvers in ball valve are recommended as much as possible to admit the air required into the system through an air valve.

Notation

A	= cross sectional area of pipe (m^2)
A_{adm}	= cross sectional area of the air valve (m^2)
$C_{d,adm}$	= inflow discharge coefficient (–)
D	= internal pipe diameter (m)
D_{av}	= air valve diameter or orifice size (mm)
f	= Darcy-Weisbach friction factor (–)
g	= gravity acceleration (ms^{-2})
h_m	= minor losses (m)
L_e	= length of emptying column (m)
L_T	= pipe length (m)
k	= polytrophic coefficient (–)
R_v	= resistance coefficient (ms^2m^{-6})
m	= mass (kg)
p_1	= pressure of air pocket (Pa)
p_{atm}	= atmospheric pressure (Pa)
Q	= discharge (m^3s^{-1})
R	= air constant (287 $Jkg^{-1}{}^{\circ}K^{-1}$)
t	= time (s)
T	= air temperature ($^{\circ}K$)
T_m	= valve maneuvering time (s)
v	= velocity (ms^{-1})
V	= volume (m^3)
x	= air pocket size (m)
z	= elevation of pipe (m)
Δz_1	= elevation difference (m)
ρ	= density (kgm^{-3})
γ	= unit weight (Nm^{-3})

Superscripts

*	= absolute values (e.g., absolute pressure)

Subscripts

0	= initial condition (e.g., initial length of emptying column)
a	= refers to air (e.g., air density)
w	= refers to water (e.g., water density)
nc	= normal conditions

Chapter 4

Subatmospheric pressure in a water draining pipeline with an air pocket

This chapter is extracted from the paper:

Subatmospheric pressure in a water draining pipeline with an air pocket
Coauthors: Coronado-Hernández O. E; Fuertes-Miquel, V. S; Besharat, M.; and Ramos, H. M.
Journal: Urban Water Journal ISSN 1573-062X.
2017 Impact Factor: 2.744. Position JCR 17/90 (Q1). Water Resources.
State: Published. 2018. Volume 15; Issue 4. DOI: 10.1080/1573062X.2018.1475578.

4.2   Introduction

Drops of absolute pressure which can cause the collapse of the system depending on the conditions of the installation may occur during the draining water process in pipelines. Emptying process is a normal procedure that engineers have to face in order to plan the operation, control, and the management of the water supply networks.

The water drainage process has been studied by a few authors in the literature. Laanearu et al. (2012) studied the behavior of the draining process in a pipeline using pressurized air in an experimental facility. In chapter 3 is developed a mathematical model for simulating the draining process to compute the main hydraulic variables in a single pipe considering two possible case: (i) corresponding to the case where an air valve has not been installed or when it has failed due to operational and maintenance problems (Ramezani et al., 2015) which represents the worse case due to causing the lowest troughs of subatmospheric pressure, and (ii) corresponding to the case where an air valve has been installed in the highest point to give reliability by admitting air into the pipe for preventing troughs of subatmospheric pressure. However, there is a lack of information regarding the behavior of a pipeline with irregular profile and without air valves. To face this problem, it is important to consider that drain valves are located at low points along pipelines and when they are opened, then the atmospheric pressure is presented downstream them. Due to not admitting air into the pipeline, the pipeline can not be completely emptied, and troughs of the subatmospheric pressure occurrence will affect not only the pipe but also some devices such as valves, joints, pumps, turbines, among others. The development of a reliability model for simulating this event, can be used for detecting problems related to subatmospheric pressure occurrence in real pipelines. In chapter 5 will implement and validate the resolution of ii) aforementioned applied to a pipeline of irregular profile with an air valve

This chapter presents a $1D$ mathematical model for simulating the draining process in a pipeline of irregular profile and without air valves. The mathematical model includes the equation for the water phase describes by the rigid water column model (Coronado-Hernández et al., 2017b; Fuertes-Miquel et al., 2016, 2018b; Izquierdo et al., 1999; Liou and Hunt, 1996), the equation for the moving air-water interface (Zhou et al., 2013a,b; Izquierdo et al., 1999; Coronado-Hernández et al., 2017b), and the equation for the air phase described by the polytropic model (Zhou et al., 2013a; Zhou and Liu, 2013; Izquierdo et al., 1999; Martins et al., 2015). The numerical resolution gives information about the air pocket pressure, the water velocity and the position of the water column. Finally, the mathematical model is validated in an experimental facility which consists of a pipeline with irregular profile of 7,3 m long, with a nominal diameter of DN 63 mm, and without air valves.

4.3   Mathematical model

Figure 4.1 presents the emptying process in a pipeline where air valves have not been installed along on it or where they are failed due to lack of maintenance and operational problems. This is the most critical situation in this procedure because it generates the lowest troughs of subatmospheric pressure which can produce the collapse of the system. Initially, the air in the pockets are at atmospheric pressure (101325 Pa). Valves CV_1 and CV_2 are closed to drain the system between them. When drain valves DV_s are opened, the air pressure decreases rapidly in the air pockets until they reach the lowest troughs of subatmospheric pressure. The main trough occurs in the first oscillation of the hydraulic event. Then, some oscillations in the absolute pressure pattern are observed. Hence, a setback occurs while emptying the water column. Finally can occur two situations: (i) the draining is stopped, and part of the water can remain inside the pipeline when blackflow air does not occur; and (ii) the draining of the system will be completed when the backflow phenomenon occurs.

A common configuration of the pipeline (see Figure 4.1) presents n air pockets located at the high points, drain valves located at the low points of the system, and k pipes in the installation. This problem is described by three main hydraulic variables: the water velocity v_j ($j = 1, 2, ..., k$), the length of the water column $L_{w,j}$ ($j = 1, 2, ..., k$), and the absolute pressure of the air pocket p_i^* ($i = 1, 2, ..., n$). The length of the pipe L_j ($j = 1, 2, ..., k$) can be estimated as $L_j = \sum L_{j,r}$ where r is the total numbers of branches of the pipe j. During this process friction factor f, and polytropic coefficient m can be considered constant (Coronado-Hernández et al., 2017b; Zhou et al., 2013a; Izquierdo et al., 1999).

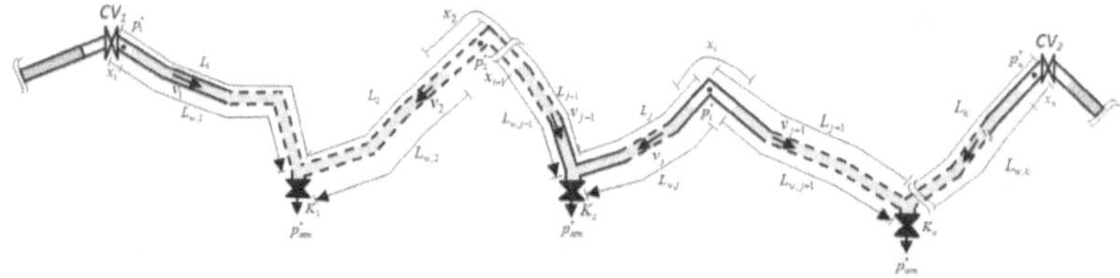

Figure 4.1: Scheme of entrapped air into a pipeline during the draining procedure

4.3.1   Governing assumptions

The one-dimensional proposed model can be applied for analyzing the emptying procedure in a pipeline with irregular profile. The proposed model has been implemented considering a previous model developed by the authors Coronado-Hernández et al. (2017b). The assumptions of the mathematical model are:

- The rigid water column model is used to simulate the water phase where fiction factor is considered constant along the hydraulic event.

- A polytropic model is used to represent the behavior of the air phase.

- The moving air-water interface is considered perpendicular to the pipe direction.

- The backflow air phenomenon does not occur during the hydraulic event, which implies that there is not air admitted by the drain valves.

4.3.2 Equations

Considering the assumptions of the mathematical model, equations for the water phase, air phase, and for the moving air-water interface can be expressed as:

Water phase equation

The water phase can be modeled by using an inertial model. The water hammer (or elastic model) considers the elastic effects of the water and the pipe. However, the elasticity of the entrapped air pocket into the pipeline is much higher than the elasticity of the water which means that the water phase can be modeled by the rigid water column model (Coronado-Hernández et al., 2017b; Izquierdo et al., 1999; Laanearu et al., 2012). Then the momentum equation can be expressed as:

$$\frac{dv_j}{dt} = \frac{p_i^* - p_{atm}^*}{\rho_w L_{w,j}} + g\frac{\Delta z_j}{L_{w,j}} - f_j\frac{v_j|v_j|}{2D} - \frac{gA^2 Q_T|Q_T|}{L_{w,j}K_s^2} \tag{4.1}$$

where p_i^* = absolute pressure of the air pocket i, ρ_w = water density, p_{atm}^* = atmospheric pressure (101325 Pa), Δz_j = elevation difference, A = cross-sectional area, Q_T = total discharge in the drain valve, and K_s = flow factor of the drain valve s.

Air phase equation

The compression and the expansion of the air pocket i obeys to the polytropic law, which relates the absolute pressure and the total volume of the air pocket:

$$p_i^* V_{a,i}^m = p_{i,0}^* V_{a,i,0}^m \tag{4.2}$$

where $V_{a,i}$ = volume of the air pocket i, $p_{i,0}^*$ = initial condition of the p_i^*, and $V_{a,i,0}$ = initial condition of the $V_{a,i}$.

However, along of the pipeline the cross-sectional area (A) is constant, then:

$$p_i^* x_i^m = p_{i,0}^* x_{i,0}^m \tag{4.3}$$

where x_i = length of the air pocket i, and $x_{i,0}$ = initial value of the x_i.

The air pocket size can be computed as $x_i = L_j - L_{w,j} + L_{j+1} - L_{w,j+1}$, thus:

$$p_i^*(L_j - L_{w,j} + L_{j+1} - L_{w,j+1})^m = p_{i,0}^*(L_j - L_{w,j,0} + L_{j+1} - L_{w,j+1,0})^m \tag{4.4}$$

Equation for the air-water interface

To compute the air-water interface, a piston flow was assumed which means that the hydraulic event occurs very fast. Then, it is perpendicular to the pipe direction where there are some reaches of the pipe completely filled by air and other by water.

$$\frac{dL_{w,j}}{dt} = -v_j \rightarrow L_{w,j} = L_{w,j,0} - \int_0^t v_j \mathrm{d}t \tag{4.5}$$

Numerical resolution

To calculate the hydraulic variables during the emptying process v_j ($j = 1, 2, ..., k$), $L_{w,j}$ ($j = 1, 2, ..., k$), and p_i^* ($i = 1, 2, ..., n$) a differential-algebraic equations (DAE) system integrated by equations 4.1, 4.4 and 4.5 has to be solved which is compound by $2k + n$ equations. The initial conditions for the system are described by: $v_j(0) = 0$, $L_{w,j}(0) = L_{w,j,0}$, and $p_i^*(0) = p_{atm}^* = 101325$ Pa.

4.4 Model verification

4.4.1 Description of experimental facility

Figure 4.2 shows the experimental facility developed at the Instituto Superior Técnico, CERIS, University of Lisbon (Portugal), where measurements were conducted to study the behavior of the draining process in a pipeline with irregular profile. The system is configured by two PVC pipes with a total length of 7,3 m and nominal diameter of 63 mm. There are two water columns represent by $L_{w,1}$ and $L_{w,2}$ which will empty by drain valves DV_1 and DV_2. The water velocity of the water columns was measured with an Ultrasonic Doppler Velocimetry (UDV) located at the horizontal PVC pipe with a transducer of 4 MHz frequency. The length of the water columns was measured by using a Sony Camera DSC-HX200V. The drain valves have a nominal diameter of 25 mm. The maneuvering time and the opening percentage valves were similar during the draining process, then the resistance coefficients are the same ($K_1 = K_2 = K_s$). The drain valves have a free discharge. There is only an air pocket inside the system which is distributed in the two PVC pipes (see detail of Figure 4.2). The air pocket size can be computed as $x_1 = 0{,}04 + x_{1,1} + x_{1,2}$. A transducer PT_1 was installed in the high point to measure the absolute pressure pattern. A pico-scope device was used to record these values.

Also, the gravity term suggested by Coronado-Hernández et al. (2017b) was included in order to consider the slope change in the water column 1:

$$\frac{\Delta z_1}{L_{w,1}} = \left(1 - \frac{L_{1,2}}{L_{w,1}}\right) \sin(30°) \tag{4.6}$$

The gravity term for the water column 2 can be computed in similar way used to the water column 1.

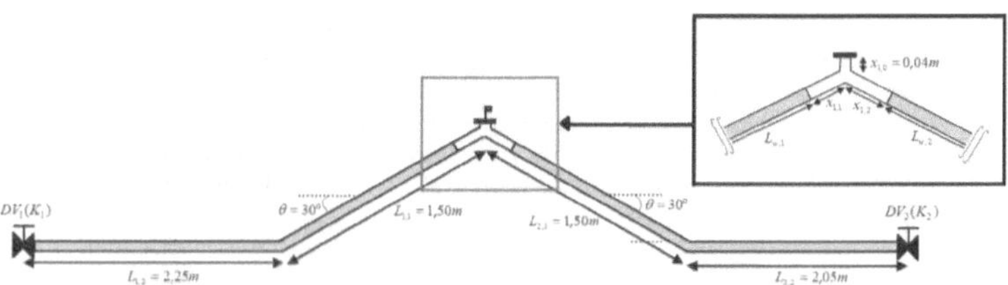

Figure 4.2: Experimental facility

4.4.2 Conceptual description of the hydraulic event and limitations of the mathematical model

The general behavior during the emptying process without air valves is presented in Fig. 4.3 which corresponds to the case where the water columns start with the air pocket distributed non-symmetrically. Figure 4.3a shows the initial conditions where the air pocket begins at atmospheric pressure ($p_{1,0}^* = p_{atm}^*$). After opening the drain valves (see Fig. 4.3b), the water column 1 presents a reverse flow moving from point $U1$ to point $U2$, whereas the water column 2 is accelerating and it moves from $U1'$ to point $U2'$ indicating the water flow is coming out of this pipe. Immediately the backflow air phenomenon starts at the outlet position with a small length. The minimum trough of the subatmospheric pressure is presented during this stage, when drain valves are totally opened. The mathematical model can not predict the backflow air phenomenon. According to the experiments, the air-water interface is practically perpendicular to the pipe direction for inclined pipes as assumption of the mathematical model. The hydraulic event continues as shown in Fig. 4.3c with the moving of the water columns in the same direction until they reach point $U3$ and point $U3'$ for water column 1 and 2, respectively. The absolute pressure pattern increases generating several oscillations. These oscillations are generated because the water velocities present oscillations around of the main direction of this movement. The length of the backflow air is greater than the previous one. Then, in Fig. 4.3d is presented the absolute pressure pattern continues changing, and the main direction of the moving of the water columns change in the opposite way from point $U3$ to $U4$ and from point $U3'$ to $U4'$ for water column 1 and 2, respectively. The backflow air in pipe 1 continues increasing and its size is divided in some parts. The backflow air starts at the outlet position in pipe 2. After that, some oscillations in the main direction regarding to the water velocities and the lengths of the water columns occur as shown in Fig 4.3e. It is very important to note that if the air pocket is distributed symmetrically in water columns, the behavior of them will be similar. Figure 4.3f shows that after some seconds of the hydraulic event, the backflow air size is increasing and moving from downstream to upstream in both water columns. The backflow air reliefs the subatmoshperic pressure pattern and produces a faster drainage of the system. In this case, the mathematical model can not predict the hydraulic behavior, and it can not be applied. However, if the sizes or the opening percentage of the drain valves are low, then backflow air does not occur, and some parts of the water columns remain into the pipeline where the draining is stopped. Finally, the air continues entering into the water columns and rapidly the draining is carried out until it reaches an open-channel flow in the pipes (see Fig. 4.3g).

4.4.3 Proposed model definition

The system presented in Fig. 4.2 has been solved with the following data: $L_1 = 3{,}77$ m, $L_2 = 3{,}57$ m, $f = 0{,}018$, $D = 51{,}4$ mm, $m = 1{,}1$, and $p_{1,0}^* = 101325$ Pa. The DAE system is given by:

1. Rigid water column model applied to the water column 1

$$\frac{dv_1}{dt} = \frac{p_1^* - p_{atm}^*}{\rho_w L_{w,1}} + g\frac{\Delta z_1}{L_{w,1}} - f\frac{v_1|v_1|}{2D} - \frac{gA^2 v_1|v_1|}{K_1^2 L_{w,1}} \tag{4.7}$$

2. Air-water interface for the water column 1

$$\frac{dL_{w,1}}{dt} = -v_1 \rightarrow L_{w,1} = L_{w,1,0} - \int_0^t v_1 dt \tag{4.8}$$

3. Rigid water column model applied to the water column 2

$$\frac{dv_2}{dt} = \frac{p_1^* - p_{atm}^*}{\rho_w L_{w,2}} + g\frac{\Delta z_2}{L_{w,2}} - f\frac{v_2|v_2|}{2D} - \frac{gA^2 v_2|v_2|}{K_2^2 L_{w,2}} \tag{4.9}$$

4. Air-water interface for the water column 2

$$\frac{dL_{w,2}}{dt} = -v_2 \rightarrow L_{w,2} = L_{w,2,0} - \int_0^t v_2 dt \tag{4.10}$$

5. Polytropic model for the air pocket 1

$$p_1^*(L_1 - L_{w,1} + L_2 - L_{w,2})^m = p_{1,0}^*(L_1 - L_{w,1,0} + L_2 - L_{w,2,0})^m \tag{4.11}$$

4.4.4 Results and discussion

There are 3 types of behavior according to the experiments. Type A corresponds to a partial opening of the drain valves by considering the air pocket volume is distributed equally inside water columns 1 and 2; Type B is similar to the aforementioned but considering a completely opening of the drain valves; and Type C considers that the air pocket volume is not distributed uniformly inside the water columns. Table 4.1 shows the runs considered during the experiments. The valve maneuvering time (T_m) for all runs was 0,7 s. For partial opening the

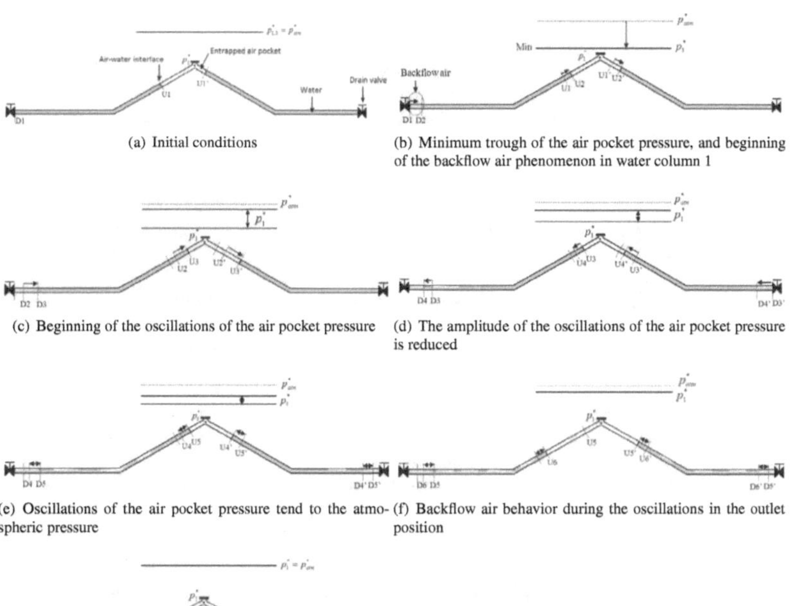

(a) Initial conditions

(b) Minimum trough of the air pocket pressure, and beginning of the backflow air phenomenon in water column 1

(c) Beginning of the oscillations of the air pocket pressure

(d) The amplitude of the oscillations of the air pocket pressure is reduced

(e) Oscillations of the air pocket pressure tend to the atmospheric pressure

(f) Backflow air behavior during the oscillations in the outlet position

(g) Final conditions

Figure 4.3: Sequence of the hydraulic events

flow factor (K_s) was $0,9 \times 10^{-4}$ m^{3}s^{-1}m$^{1/2}$, whereas for the total opening it was $1,4 \times 10^{-3}$ m^{3}s^{-1}m$^{1/2}$.

Table 4.1: Types of behaviour

Type	Run No.	Opening drain valves (%)	$x_{1,1}$ (m)	$x_{1,2}$ (m)	x_1 (m)
A	1	Partial	0,28	0,28	0,60
A	2	Partial	0,61	0,61	1,26
B	3	Completely	0,28	0,28	0,60
B	4	Completely	0,61	0,61	1,26
C	5	Completely	0,26	0,60	0,90
C	6	Completely	0,28	0,98	1,30

Type A: Partial opening of the drain valves with air pocket volume distributed uniformly

Figure 4.4 shows the results which corresponds to the less critical case for the pipeline because the subatmospheric pressure pattern reaches values close to the atmospheric pressure head. The hydraulic event starts with the atmosphetic pressure head (10,33 m), and after decreases rapidly until reaches the trough of the subatmospheric pressure head of 9,74 m and 9,90 for run No. 1 and No. 2, respectively. Practically there are not oscillations on the evolution of the absolute pressure pattern during the first two seconds, as a consequence of the partial opening of the ball valves. After this time, the subatmospheric pressure pattern remains constant. The mathematical model shows an excellent fit regarding the two experiments in each run. For Type A was not possible to measure the water velocities because during the hydraulic event they reached values lower than 0,015 m/s which can not be detected by the UDV.

Type B: Total opening of the drain valves with air pocket volume distributed uniformly

The initial conditions of the Type B are similar to the Type A, but now the two ball valves are opened completely. As a consequence, the troughs of the subatmospheric pressure reached in the Type B are lower than in the Type A which indicates that the risk of collapse for the Type B is higher than for the Type A.

Again, the subatmospheric pressure pattern in the hydraulic event starts in atmospheric conditions, and after decreases quickly until reached the minimum values of 9,28 m (at 0,4 s) and 9,65 (at 0,5 s) for runs No. 3 and No. 4, respectively (see Figure 4.5). Then, some oscillations are presented along of the hydraulic event which indicates the subatmospheric pressure pattern is able to move from upstream to downstream and reciprocally the two water columns behave as a piston flow. The mathematical model predicted adequately runs No. 3 and No. 4.

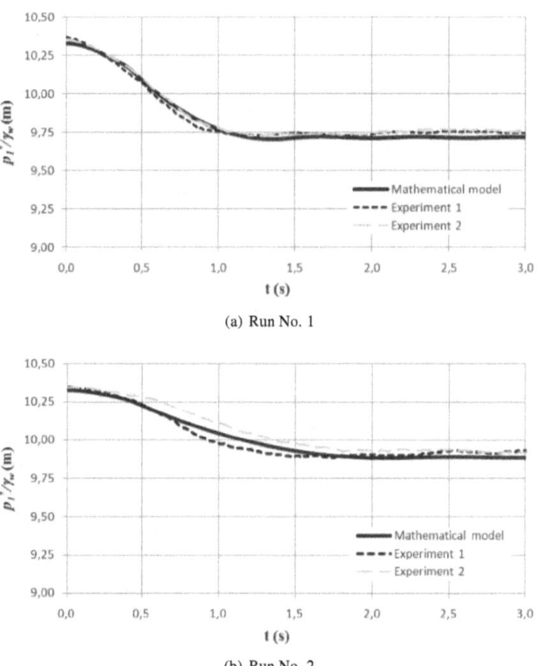

(a) Run No. 1

(b) Run No. 2

Figure 4.4: Comparison between computed and experiments of the absolute pressure pattern for Type A

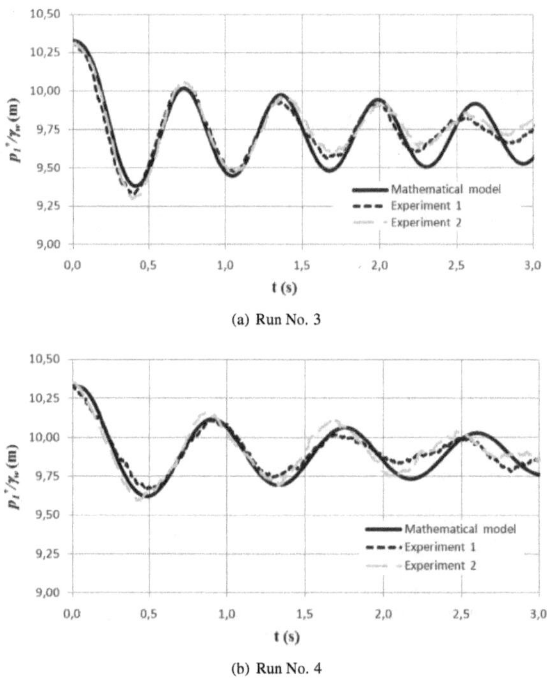

(a) Run No. 3

(b) Run No. 4

Figure 4.5: Comparison between computed and experiments of the absolute pressure pattern for Type B

Figure 4.6 shows the results concerning to the water velocities. For runs No. 3 and No. 4 the values of the water velocities in the two water columns are similar due to the air pocket size is distributed uniformly. For both runs, the water velocity increases rapidly until reaching a maximum value. Subsequently, the water velocity decreases until it reaches a value of 0 m/s at 0,4 s and 0,49 s for runs No. 3 and No. 4, respectively. Then it continues to decrease until reaching a minimum value of the $-0,11$ m/s for run No. 3 and $-0,125$ m/s for run No. 4, indicating negative velocities occurrence. After some oscillations, with a similar amplitude, a setback of the water columns occur. The mathematical model can follow the behavior of the water velocity experiments.

Type C: Total opening of the drain valves with air pocket volume distributed non-uniformly

Type C is the most complex to be analyzed because the behavior of the two water columns is so different. Figure 4.7a shows the results of the subatmospheric pressure pattern for run No. 5 where according to the experiments the trough of the subatmospheric pressure presented is 9,50 m at 0,4 s. The two water columns have different movements, then the amplitude of the oscillations of the air pocket along the transient is lower compared with the Type B. The mathematical model predicts accurately the subatmospheric pressure pattern along of the hydraulic event. However, Figure 4.7b presents the results for run No. 6 where the mathematical model can predict only the first oscillation during the transient. This case occurs because the mathematical model considers an air-water interface perpendicular to pipe walls at downstream of the water columns, which does not really happen for an air pocket distributed asymmetrically.

Figure 4.8 presents the results of the water velocities for two water columns for run No. 5 which are different due to subatmospheric pressure pattern. Figure 4.8a shows the behaviour of the water column 1 where velocity values always are positive generating that in this column the emptying process occurs. According to the experiments, the maximum velocity is presented at 1,05 s with a value of 0,175 m/s, which is very close to the value predicted by the mathematical model (0,19 m/s). In contrast, Figure 4.8b presents the results for the water column 2 where according to the experiments negative velocities occur. As a consequence, the water column 2 moves from downstream to upstream during the hydraulic event. The mathematical models predicts the tendency regarding water velocities for both water columns.

Type C represents the case where the evolution of the lengths of the water columns 1 and 2 are different during the transient. Figure 4.9 shows that the water column 1 has at the beginning of the transient an initial value of 3,5 m and at the end of the hydraulic event a value of 3,2 m, indicating that 0,3 m could be drained during the process. In contrast, the water column 2 starts with 3,0 m (at 0 s) and finishes with 3,2 m (at 3 s), indicating that the entire water column 2 was displaced 0,2 m from downstream to upstream and practically this water column could not be drained. The mathematical model predicts accurately the length of the water columns.

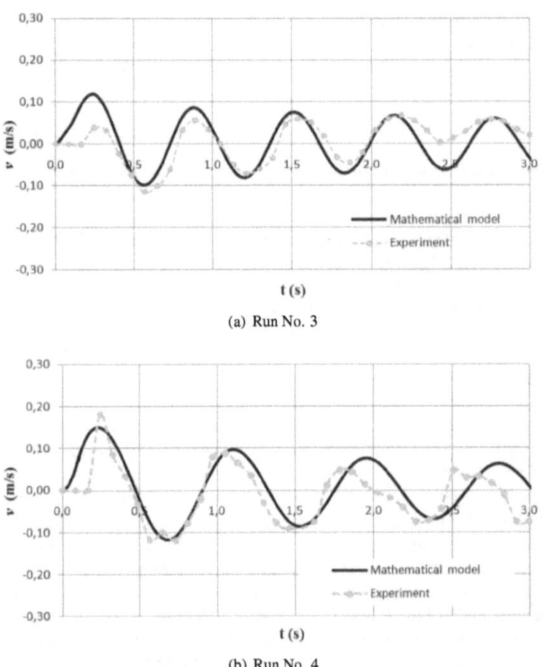

(a) Run No. 3

(b) Run No. 4

Figure 4.6: Comparison between computed and experiments of the water velocity for Type B

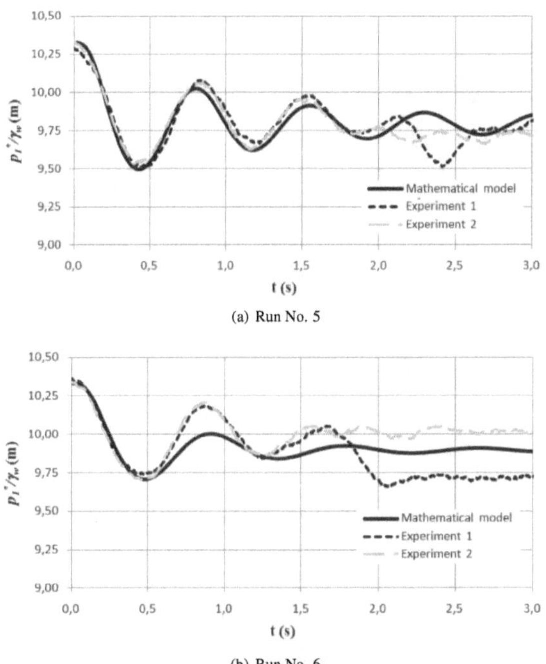

(a) Run No. 5

(b) Run No. 6

Figure 4.7: Comparison between computed and experiments of the absolute pressure pattern for Type C

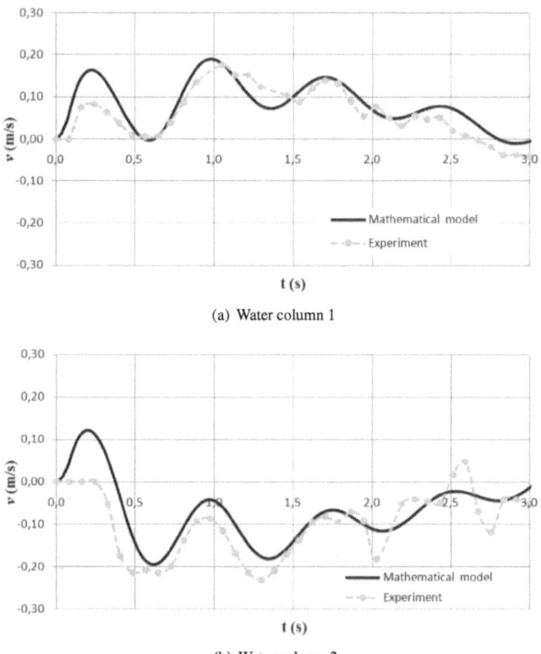

(a) Water column 1

(b) Water column 2

Figure 4.8: Comparison between computed and experiments of the water velocities for Type C and run No. 5

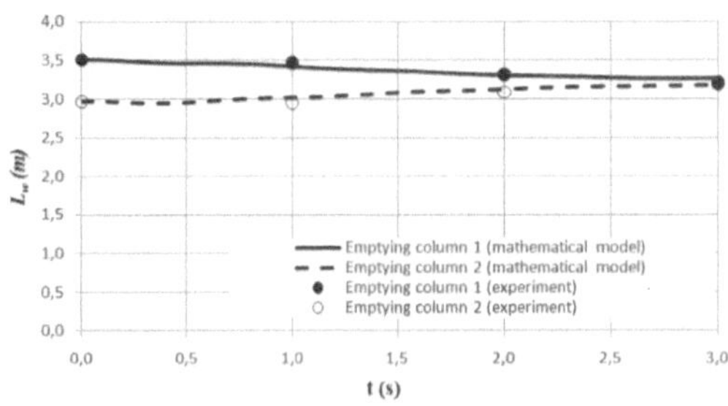

Figure 4.9: Comparison between computed and experiments of the length of water columns for Type C (run No. 5)

4.5 Case study and results

4.5.1 Description of case study

The proposed model is used to simulate the emptying process of a pipeline of irregular profile with four branches and three air pockets. The system has the following data: $D = 0{,}3$ m, $L_1 = 700$ m (with length branches $L_{1,1}$ and $L_{1,2}$, and slope branches $\theta_{1,1}$ and $\theta_{1,2}$), $L_2 = 600$ m (with slope branch θ_2), $L_3 = 600$ m (with slope branch θ_3), and $L_4 = 700$ m (with length branches $L_{4,1}$, $L_{4,2}$, and $L_{4,3}$, and slope branches $\theta_{4,1}$, $-\theta_{4,2}$, and $\theta_{4,3}$). The drainage of the system is carried out by drain valves DV_1 and DV_2 with flow resistance coefficient values of K_1 and K_2, respectively. The air pocket sizes are x_1, x_2, and x_3 with absolute pressure values of p_1^*, p_2^*, and p_3^*, respectively. Figure 4.10 shows characteristics of the installation.

4.5.2 Proposed model definition

The corresponding equations of the pipeline are:

1. Mass oscillation equation applied to the emptying column 1

$$\frac{dv_1}{dt} = \frac{p_1^* - p_{atm}^*}{\rho_w L_{w,1}} + g\frac{\Delta z_1}{L_{w,1}} - f\frac{v_1|v_1|}{2D} - \frac{gA^2(v_1 + v_2)|v_1 + v_2|}{K_1^2 L_{w,1}} \qquad (4.12)$$

2. Air-water interface for the water column 1

$$\frac{dL_{w,1}}{dt} = -v_1 \rightarrow L_{w,1} = L_{w,1,0} - \int_0^t v_1 dt \qquad (4.13)$$

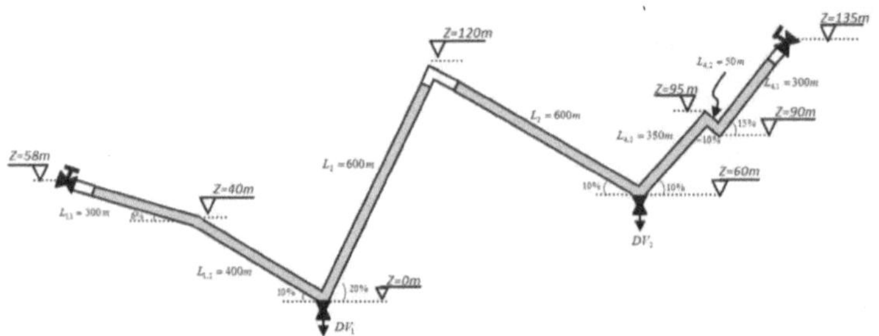

Figure 4.10: Pipeline of irregular profile without air valves

3. Mass oscillation equation applied to the emptying column 2

$$\frac{dv_2}{dt} = \frac{p_2^* - p_{atm}^*}{\rho_w L_{w,2}} + g\frac{\Delta z_2}{L_{w,2}} - f\frac{v_2|v_2|}{2D} - \frac{gA^2(v_1 + v_2)|v_1 + v_2|}{K_1^2 L_{w,2}} \qquad (4.14)$$

4. Air-water interface for the water column 2

$$\frac{dL_{w,2}}{dt} = -v_2 \rightarrow L_{w,2} = L_{w,2,0} - \int_0^t v_2 dt \qquad (4.15)$$

5. Mass oscillation equation applied to the emptying column 3

$$\frac{dv_3}{dt} = \frac{p_3^* - p_{atm}^*}{\rho_w L_{w,3}} + g\frac{\Delta z_3}{L_{w,3}} - f\frac{v_3|v_3|}{2D} - \frac{gA^2(v_3 + v_4)|v_3 + v_4|}{K_2^2 L_{w,3}} \qquad (4.16)$$

6. Air-water interface for the water column 3

$$\frac{dL_{w,3}}{dt} = -v_3 \rightarrow L_{w,3} = L_{w,3,0} - \int_0^t v_3 dt \qquad (4.17)$$

7. Mass oscillation equation applied to the emptying column 4

$$\frac{dv_4}{dt} = \frac{p_3^* - p_{atm}^*}{\rho_w L_{w,4}} + g\frac{\Delta z_4}{L_{w,4}} - f\frac{v_4|v_4|}{2D} - \frac{gA^2(v_3 + v_4)|v_3 + v_4|}{K_2^2 L_{w,4}} \qquad (4.18)$$

8. Air-water interface for the water column 4

$$\frac{dL_{w,4}}{dt} = -v_4 \rightarrow L_{w,4} = L_{w,4,0} - \int_0^t v_4 \mathrm{d}t \tag{4.19}$$

9. Polytropic model for the air pocket 1

$$p_1^* x_1^m = p_{1,0}^* x_{1,0}^m \tag{4.20}$$

where $x_1 = L_1 - L_{w,1}$.

10. Polytropic model for the air pocket 2

$$p_2^* x_2^m = p_{2,0}^* x_{2,0}^m \tag{4.21}$$

where $x_2 = L_2 - L_{w,2} + L_3 - L_{w,3}$.

11. Polytropic model for the air pocket 3

$$p_3^* x_3^m = p_{3,0}^* x_{3,0}^m \tag{4.22}$$

where $x_3 = L_3 - L_{w,3}$.

A 11x11 system of DAE describe the whole problem. Together with the corresponding boundary and initial conditions, 11 unknowns can be solved v_1, $L_{w,1}$, v_2, $L_{w,2}$, v_3, $L_{w,3}$, v_4, $L_{w,4}$, p_1^*, p_2^*, and p_3^*. The initial conditions at $t = 0$ are described by $v_1(0) = 0$, $L_{w,1}(0) = L_{w,1,0}$, $v_2(0) = 0$, $L_{w,2}(0) = L_{w,2,0}$, $v_3(0) = 0$, $L_{w,3}(0) = L_{w,3,0}$, $v_4(0) = 0$, $L_{w,4}(0) = L_{w,4,0}$, $p_1^*(0) = p_{atm}^*$, $p_2^*(0) = p_{atm}^*$, and $p_3^*(0) = p_{atm}^*$. The upstream boundary conditions are given by $p_{1,0}^*$, $p_{2,0}^*$, and $p_{3,0}^*$ (initial conditions of air pockets), and the downstream boundary condition are given by p_{atm}^* (water free discharge to the atmosphere). Gravity terms $\Delta z_1 / L_{w,1}$, $\Delta z_2 / L_{w,2}$, $\Delta z_3 / L_{w,3}$, and $\Delta z_4 / L_{w,4}$ can be computed as presented by Coronado-Hernández et al. (2017b).

4.5.3 Results

Figure 4.11 shows results for the pipeline of irregular profile without air valves for the main hydraulic and thermodynamic variables along the transient event (absolute pressure, water velocity, and length of the water column). The initial air pocket sizes were $x_{1,0} = x_{2,0} = x_{3,0} = 250$ m. A polytropic coefficient (k) of 1,2, and a friction factor (f) of 0,018 were considered.

According to Figure 4.11, the authors can deduce:

- At the beginning of the transient, maximum water velocities are reached in water columns as shown in Figure 4.11a. Afterwards, they decrease rapidly until they reach their minimum values. At the end, some oscillations occur in the water velocity. The

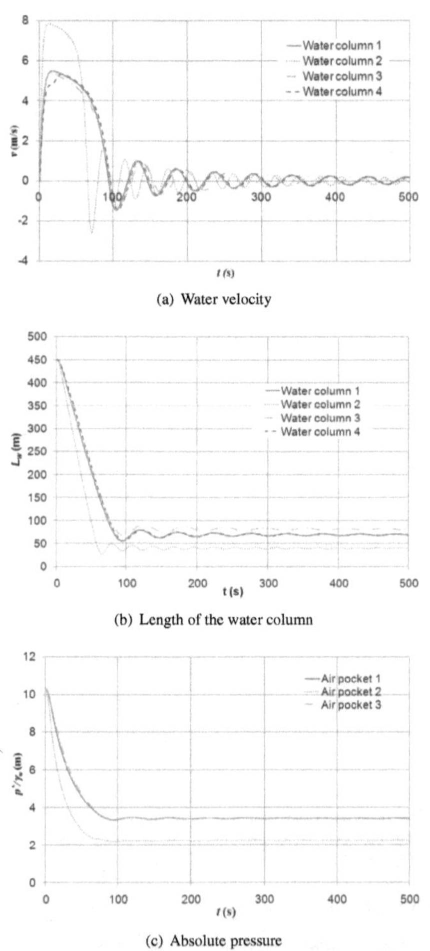

(a) Water velocity

(b) Length of the water column

(c) Absolute pressure

Figure 4.11: Full transient analysis of the pipeline of irregular profile without air valves

maximum water velocity is reached in water column 1 when it abandons the branch $L_{1,1}$ (300 m long.) and stars the branch $L_{1,2}$ (400 m long.) at 16,63 m with a value of 5,49 ms^{-1} (388,5 ls^{-1}). For water columns 2, 3, and 4, the maximum water velocities are 7,86 ms^{-1}, 5,39 ms^{-1}, and 5,34 ms^{-1}, respectively. The minimum water velocities are: for water column 1, $-1,47$ ms^{-1}; for water column 2, $-2,63$ ms^{-1}; for water column 3, $-1,34$ ms^{-1}; and for water column 4, $-1,47$ ms^{-1}.

- The lengths of the water columns start to be drained linearly until setback occurs when draining is stopped, and part of the water columns remain in the pipeline (see Figure 4.11b). If backflow phenomenon does not occur the pipeline cannot drain completely, and part of the water columns can remain in the installation with values of 70 m, 41 m, 81 m, and 70 m for water columns 1, 2, 3, and 4, respectively.

- Initially, the air pockets are at atmospheric pressure (101325 Pa). The air pressure heads decrease rapidly in the pockets until reach the minimum values of 3,31 m, 2,18 m, and 3,32 m, for the air pockets 1, 2, and 3, respectively, at 93 s (see Figure 4.11c). The backflow phenomenon does not affect the minimum values of the subatmospheric pressure.

4.6   Conclusions

The mathematical model developed by the authors predicts adequately the behavior of the draining process for the first seconds of the transient. This is very important since the minimum value of the subatmospheric pressure occurs in the first oscillation of the hydraulic event. To validate the mathematical model, several experiments were carried out in an experimental facility where the comparison between the computed and the measured of the main hydraulic variables (absolute pressure of the air pocket, water velocity and the length of the water column) confirms the goodness of the mathematical model. Regarding the results, the following conclusions can be drawn:

- Three types of behavior have been analyzed. They depend on the opening percentage of the drain valves, and the distribution of the air pocket volume inside the water columns.

- The backflow air phenomenon depends on the opening percentage of the drain valves: (i) for partial opening, it did not occur during the experiments (Type A), and (ii) for total opening, the air moved from downstream to upstream (Types B and C).

- The Type C is the most complex case to predict by the mathematical model. When the initial difference elevation (Δz_j) is very close for the water columns (run No. 5), then the mathematical model can predict the evolution of the hydraulic variables. However, when the difference elevation is high (run No. 6), then the mathematical model can only predict the first oscillation during the hydraulic event.

The mathematical model cannot predict the backflow air phenomenon. However, it reduces the risk of collapse the system by introducing air at atmospheric pressure into the pipeline. As a consequence, the backflow reliefs the troughs of the subatmospheric pressure, and rapidly the system can reach the atmospheric conditions.

Notation

A	= cross sectional area of pipe (m^2)
D	= internal pipe diameter (m)
f	= Darcy-Weisbach friction factor (–)
g	= gravity acceleration (ms^{-2})
K_s	= flow factor of the drain valve s (m^{3}s^{-1}m$^{1/2}$)
$L_{w,j}$	= length of the water column j (m)
L_j	= total length of the pipe j (m)
m	= polytrophic coefficient (–)
p_i	= pressure of the air pocket i (Pa)
p_{atm}	= atmospheric pressure (Pa)
Q_T	= total discharge (m^{3}s^{-1})
t	= time (s)
T_m	= valve maneuvering time (s)
r	= number of reaches of the pipe j (–)
$V_{a,i}$	= volume of the air pocket i (m^3)
v_j	= water velocity of the water column j (ms^{-1})
x_i	= length of the air pocket i (m)
Δz_j	= elevation difference of the water column j (m)
ρ	= density (kgm^{-3})
DV_s	= Drain valve s
CV_h	= Used valves to establish the boundary conditions

Superscripts

*	= absolute values (e.g., absolute pressure)

Subscripts

0	= initial condition (e.g., initial length of the water column)
w	= refers to water (e.g., water density)
a	= refers to air pocket (e.g., air pocket pressure)

Chapter 5

Experimental and numerical analysis of a water emptying pipeline using different air valves

This chapter is extracted from the paper:

Experimental and numerical analysis of a water emptying pipeline using different air valves
Coauthors: Coronado-Hernández O. E; Fuertes-Miquel, V. S; Besharat, M.; and Ramos, H. M.
Journal: Water ISSN 2073-4441.
2017 Impact Factor: 2.069. Position JCR 34/90 (Q2). Water Resources.
State: Published. 2017. Volume 9; Issue 2; 98. DOI: 10.3390/w9020098.

5.2 Introduction

The emptying procedure in a pipeline generates hydraulic events which can cause problems if air valves are not well sized. In practical applications, engineers follow typical recommendations from the American Water Works Association (AWWA) (AWWA, 2001) or manufacturers about sizing and location of air valves along a pipeline in order to avoid subatmospheric pressure that can cause the collapse of the system. It is recommended in a pipeline that the air volume admitted by air valves should be the same as the water volume drained (Ramezani et al., 2015), consequently air valves should work in subsonic flow condition. Air valves should be located at high points, in long horizontal pipe branches, long descents, long ascents, decrease in an up-slope and at increase in a down-slope of a pipeline, and on the discharge side of deep well pumps and at vertical turbines/pumps (AWWA, 2001). An inappropriate selection of the air valve size and location produces not only subatmospheric pressure but also a slow drainage of the system.

Presently there are only few studies related to the emptying process in the literature, but they are not focused on practical applications because they do not consider a pipeline with an irregular profile and air valves (Tijsseling et al., 2016; Laanearu et al., 2012).

The analysis of an emptying process is not trivial because it requires the study of a two phase flow (Fuertes, 2001; Besharat et al., 2016; Apollonio et al., 2016; Balacco et al., 2015). This problem can be studied using one $(1D)$ (Zhou et al., 2013a; Izquierdo et al., 1999), two (2D) (Zhou et al., 2011b) or three-dimensional (3D) (Martins et al., 2016) models. The water phase in 1D model can be analyzed considering two types of models (Abreu et al., 1999): (i) elastic models (Martins et al., 2010c,a), which consider the elasticity of the pipe and the water, or (ii) rigid models (Fuertes-Miquel et al., 2016), which ignore the elasticity of them. Normally, elastic models are solved by using the method of characteristics (Zhou and Liu, 2013; Covas et al., 2010), and rigid models by using the numerical solution of ordinary differential equations (Izquierdo et al., 1999; Fuertes-Miquel et al., 2016; Liou and Hunt, 1996). In pressurized systems the air effect can be analyzed as a single-phase flow, where the absolute pressure of the air pocket is computed between two water columns (Bousso et al., 2013; Zhou et al., 2013a; Izquierdo et al., 1999). Regarding the analysis of 2D and 3D CFD modeling of air-water interface in closed pipes, they are still unyielding in the application of pipeline draining because length and time scales are not appropriate.

This research develops a general 1D mathematical model that can be used for analyzing the behavior of the main hydraulic variables during the emptying process in a pipeline with irregular profile and with several air valves installed along on it based on formulations of previous works (Izquierdo et al., 1999; Zhou et al., 2013a; Liou and Hunt, 1996; Leon et al., 2010; Wylie and Streeter, 1993; Fuertes-Miquel et al., 2016, 2018b). The proposed model can give important information in real systems about: (i) the risk of collapse of pipeline installations, by checking in a pipe manufacturer characteristics if the stiffness class is appropriate to support the subatmospheric pressure reached during the hydraulic event depending on the type of soil in natural conditions, the type of backfill and the cover depth; (ii) the appropriate selection of the air valve during the emptying process; (iii) the size and the maneuver time of

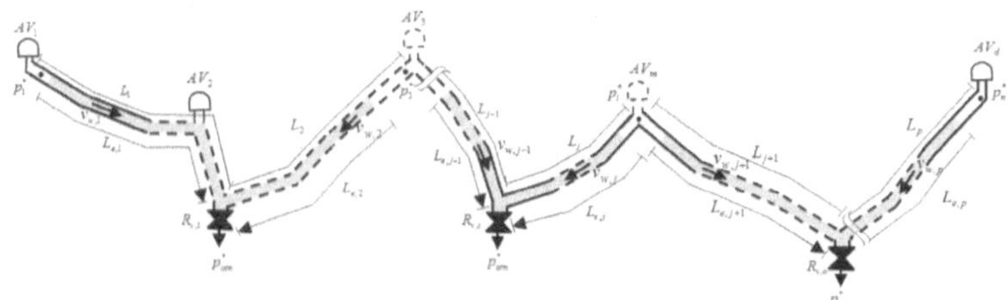

Figure 5.1: Scheme of entrapped air pockets in a pipeline with irregular profile while water empties

the drain valves; and (iv) the estimation of the drainage time of a pipeline.

5.3 Mathematical model

Figure 5.1 shows a typical configuration of an irregular profile in a pipeline which consists of n entrapped air pockets, d air valves, p pipes, and o drain valves located at low points for draining the system, being D is the internal pipe diameter (m); A is the cross section area of the pipe (m^2); f is the Darcy-Weisbach friction factor; g is the gravity acceleration (ms^{-2}). L_j $(j = 1, 2, ..., p)$ is the total length of each pipe, AV_m $(m = 1, 2, ..., d)$ are the air valves, and $R_{v,s}$ $(s = 1, 2, ..., o)$ represents the resistance coefficient for each drain valve. The emptying process starts when the drain valves are opened, thus air valves start to admit air into the pipeline. After that, the air pocket i will begin to expand generating subatmospheric pressure. At the same time, the drainage of the water column starts until the entire pipeline is completely emptied. Figure 5.1 gives the evolution at time t, the length of the emptying column $L_{e,j}$ $(j = 1, 2, ..., p)$, the absolute pressure of the air pocket p_i^* $(i = 1, 2, ..., n)$, and the water velocity $v_{w,j}$ $(j = 1, 2, ..., p)$. The expansion of the air pocket i can be computed as $x_i = L_j - L_{e,j} + L_{j+1} - L_{e,j+1}$.

The one-dimensional (1D) proposed model has the following assumptions: (1) water column has been modeled by the rigid model, (2) the Darcy-Weisbach equation was considered to evaluate friction losses, (3) the thermodynamic behavior of the air pocket is analyzed using a polytropic model, and (4) the air - water interface is perpendicular to the pipe direction. The proposed model can be used for pipelines with small diameters and with hydraulic slopes enough to prevent downstream air intrusion where open-channel flow does not occur (Liou and Hunt, 1996; Vasconcelos and Wright, 2008).

Under these hypotheses, the problem is modeled by the following set of equations.

5.3.1 Equations for the water phase

- Mass oscillation equation:

  The rigid model was used in order to compute the evolution of the water column (Cabrera et al., 1992; Izquierdo et al., 1999) considering that the elasticity of the air is much higher than the elasticity of the pipe and the water (Zhou et al., 2011b). Applying the rigid model to the emptying column j and considering that the drain valve s joins pipes L_j and L_{j-1}, which is a common point in a pipeline, then:

$$\frac{dv_{w,j}}{dt} = \frac{p_i^* - p_{atm}^*}{\rho_w L_{e,j}} + g\frac{\Delta z_{e,j}}{L_{e,j}} - f\frac{v_{w,j}|v_{w,j}|}{2D} - \frac{R_{v,s}gA^2(v_{w,j} + v_{w,j-1})|v_{w,j} + v_{w,j-1}|}{L_{e,j}}$$

(5.1)

  where $\Delta z_{e,j}$ is the elevation difference (m), ρ_w is the water density ($\mathrm{kgm^{-3}}$), p_{atm}^* is the atmospheric pressure (101325 Pa) and g is the gravity acceleration ($\mathrm{ms^{-2}}$).

  The expression $h_{m,s} = R_{v,s}Q_{w,s}^2$ was used to estimate the local loss of the drain valve s in equation 5.1. $R_{v,s}$ is the resistance coefficient and $Q_{w,s}$ is the water discharge.

  If the drain valve only connects pipe L_j, thus $v_{w,j-1} = 0$.

- Air-water interface position:

  The position of the air-water interface is considered perpendicular to the pipe direction (Zhou et al., 2013b; Zhou and Liu, 2013; Izquierdo et al., 1999). The continuity equation for the moving air-water interface j is:

$$\frac{dL_{e,j}}{dt} = -v_{w,j} \rightarrow L_{e,j} = L_{e,j,0} - \int_0^t v_{w,j}\,\mathrm{dt}$$

(5.2)

where subscript 0 refers to the initial condition.

5.3.2 Equations for air pockets

- Continuity equation (Fuertes-Miquel et al., 2016, 2018b):

$$\frac{dm_{a,i}}{dt} = \rho_{a,nc}v_{a,nc,m}A_{adm,m}$$

(5.3)

and by deriving:

$$\frac{dm_{a,i}}{dt} = \frac{d(\rho_{a,i}V_{a,i})}{dt} = \frac{d\rho_{a,i}}{dt}V_{a,i} + \frac{dV_{a,i}}{dt}\rho_{a,i}$$

(5.4)

where $m_{a,i}$ is the air mass and $V_{a,i}$ is the air volume of the air pocket i.

Due to the air pocket i located between pipes L_j and L_{j+1} then $V_{a,i} = (L_j - L_{e,j})A + (L_{j+1} - L_{e,j+1})A$ and $dV_{a,i}/dt = -(dL_{e,j}/dt)A - (dL_{e,j+1}/dt)A = A(v_{w,j} + v_{w,j+1})$, thus:

$$\frac{d\rho_{a,i}}{dt} = \frac{\rho_{a,nc}v_{a,nc,m}A_{adm,m} - (v_{w,j+1} + v_{w,j})A\rho_{a,i}}{A(L_j - L_{e,j} + L_{j+1} - L_{e,j+1})} \tag{5.5}$$

where $\rho_{a,i}$ is the air density of the air pocket i, $\rho_{a,nc}$ is the air density in normal conditions (1,205 kgm^{-3}), $A_{adm,m}$ is the cross section area (m^2) of the air valve m and $v_{a,nc,m}$ is the air velocity in normal conditions admitted by the air valve m.

- Expansion equation for the air pocket i:

The thermodynamic behavior of the air pocket (Martins et al., 2015) is treated by using a polytropic model (Martin, 1976; Leon et al., 2010).

$$\frac{dp_i^*}{dt} = -k\frac{p_i^*}{V_{a,i}}\frac{dV_{a,i}}{dt} + \frac{p_i^*}{V_{a,i}}\frac{k}{\rho_{a,i}}\frac{dm_{a,i}}{dt} \tag{5.6}$$

Considering the equations presented before, then:

$$\frac{dp_i^*}{dt} = \frac{kp_i^*}{A(L_j - L_{e,j} + L_{j+1} - L_{e,j+1})}\left(\frac{\rho_{a,nc}v_{a,nc,i}A_{adm,m}}{\rho_{a,i}} - A(v_{w,j+1} + v_{w,j})\right) \tag{5.7}$$

where k is the polytropic coefficient: if $k = 1,0$ then the process is isothermal, but if $k = 1,4$ the process is adiabatic.

In equations 5.5 and 5.7 if the air pocket i is only located in pipe L_j, then $v_{e,j+1} = 0$ and $L_{j+1} = L_{e,j+1} = 0$.

- Air valve characterization:

The formulation proposed by Wylie and Streeter (Wylie and Streeter, 1993; Iglesias-Rey et al., 2014) was used to represent the air admission into the system for air valves. Ideally the air valve m should be working in subsonic flow ($p_{atm}^* > p_i^* > 0,528p_{atm}^*$), then:

$$Q_{a,nc,m} = C_{adm,m}A_{adm,m}\sqrt{7p_{atm}^*\rho_{a,nc}\left[\left(\frac{p_i^*}{p_{atm}^*}\right)^{1,4286} - \left(\frac{p_i^*}{p_{atm}^*}\right)^{1,714}\right]} \tag{5.8}$$

where $Q_{a,nc,m}$ is the air discharge in normal conditions admitted by the air valve m and $Q_{a,nc,m} = v_{a,nc,m}A_{adm,m}$.

When air valves are not located in high points of the system, then the position of them
should be specified. When the air-water interface reaches an air valve, then it starts to admit
air inside the system.

In summary, a set of $2p + 2n + d$ equations describes the whole system. Together with
the corresponding boundary conditions, it can be solved for the $2p + 2n + d$ unknowns $v_{w,j}$,
$L_{e,j}$, p_i^*, $\rho_{a,i}$, $v_{a,nc,m}$ $(j = 1, 2, ..., p; i = 1, 2, ..., n; m = 1, 2, ..., d)$.

5.3.3 Initial and boundary conditions

When the pipeline is at rest, the initial conditions are described as follow: $v_{w,j}(0) = 0 (j =
1, 2, .., p)$, $L_{e,j}(0) = L_{e,j,0}(j = 1, 2, .., p)$, $p_i^*(0) = p_{i,0}^*(i = 1, 2, .., n)$, $\rho_{a,i}(0) = \rho_{a,i,0}(i =
1, 2, .., n)$, $v_{a,nc,m}(0) = 0(m = 1, 2, .., d)$.

The upstream boundary condition is given by $p_i^* = p_{i,0}^*$ and the downstream is given by
the resistance coefficient $R_{v,s}$ of the drain valve s, and the atmospheric pressure p_{atm}^* due to
the free discharge.

5.3.4 Gravity term

Figure 5.2 describes the evolution of the gravity term (see equation 5.1) along the emptying
column j. $L_{j,r}$ $(r = 1, 2, ..., h)$ is the length of the pipe reach r, and L_j is the total pipe
length j $(L_j = \sum_{r=1}^{r=h} L_{j,r})$. Subscript u is used to identify the pipe reaches (1 to h) where
the air-water interface is located.

The gravity term of the emptying column j is computed by:

- When the air-water interface has not arrived the last reach $(r \neq h)$:

$$\frac{\Delta z_{e,j}}{L_{e,j}} = \frac{\sum_{r=u+1}^{r=h} L_{j,r} \sin(\theta_{j,r})}{L_{e,j}} + \left(1 - \frac{\sum_{r=u+1}^{r=h} L_{j,r}}{L_{e,j}}\right) \sin(\theta_{j,u}) \qquad (5.9)$$

- When the air-water interface is located on the last reach $(r = h)$:

$$\frac{\Delta z_{e,j}}{L_{e,j}} = \sin(\theta_{j,h}) \qquad (5.10)$$

5.4 Model verification

5.4.1 Experimental model

In order to study the emptying process in a pipeline an experimental facility was developed
(see Figure 5.3) at the Civil Engineering, Research and Innovation for Sustainability (CEris)
Center, in the Hydraulic Lab of Instituto Superior Técnico (IST), University of Lisbon, Portugal. The experimental facility consisted of a set of transparent PVC pipes with 7,3 m length,
and nominal diameter of 63 mm $(DN63)$. An air valve (AV_1) is installed at the highest point

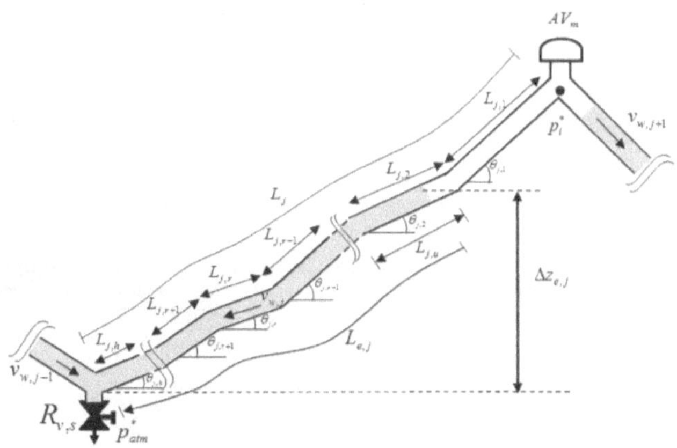

Figure 5.2: Evolution of the gravity term of the emptying column j

of the pipeline with a pressure transducer (PT_1) to measure the absolute pressure. The air valves $S050$ and $D040$ (manufacturer A.R.I.) and different air pocket sizes were tested for showing the effect in the hydraulic behavior. There are four ball valves (BV_s). BV_1, BV_2 and BV_4 are opened, consequently permit the movement of the water column. BV_3 and manual valve (MV_1) are closed and represent the system configuration extremities. The two manual ball valves (MBV_s) identified as MBV_1 and MBV_2 with nominal diameter of 25 mm $(DN25)$ at the downstream ends were used to control the outflow conditions. These valves have the same level as the horizontal pipes $L_{1,2}$ and $L_{2,2}$. Two free-surface small tanks were used to collect the drainage water. The PicoScope system was used for absolute pressure data recording. The frequency of the pressure data collection was 0,0062 s. The length of the emptying columns were measured by using a Sony Camera DSC-HX200V for decomposing frames for each second. The water velocities were measured with the Ultrasonic Doppler Velocimetry (UDV) device with a transducer for 4 MHz frequency. The transducer was located on the horizontal pipe with an angle of 20°. To measure the water velocities all other facilities were turned off in the Hydraulic Lab to avoid the noise and seeding was used inside the water in order to get appropriate measurements.

The emptying process of the experimental facility was started by an opening maneuver at the same time of valves (MBV_1) and (MBV_2). Consequently, the two emptying columns start the emptying procedure until reaching the horizontal pipes when practically the drainage is stopped, and part of the two water columns remain inside of the installation because the gravity term is zero in both pipe reaches.

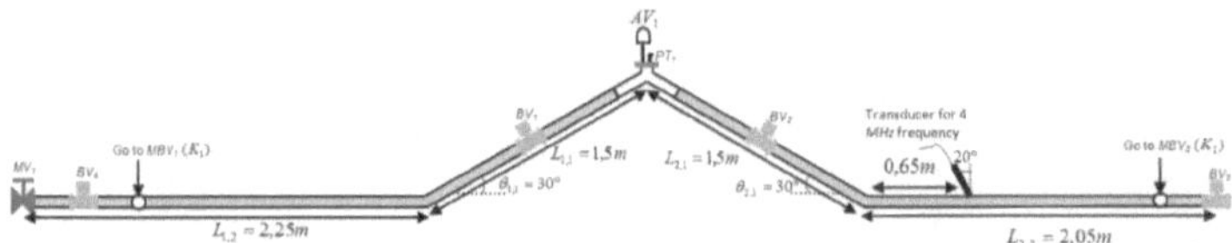

Figure 5.3: The pipe system and its components

Equations 5.13 to 5.19 were used to simulate the emptying process of this experimental facility. The gravity terms were computed for the two emptying columns. For the emptying column 1, it depends on:

- If the air-water interface is located on the sloped pipe reach $\theta = 30°$ ($L_{e,1} \leq L_{1,1} + L_{1,2}$ and $L_{e,1} > L_{1,2}$) then:

$$\frac{\Delta z_{e,1}}{L_{e,1}} = \left(1 - \frac{L_{1,2}}{L_{e,1}}\right)\sin(\theta_{1,1}) \tag{5.11}$$

- If the air-water interface is located on the horizontal pipe reach $\theta = 0°$ ($L_{e,1} > 0$ and $L_{e,1} \leq L_{1,2}$) then:

$$\frac{\Delta z_{e,1}}{L_{e,1}} = 0 \tag{5.12}$$

The gravity term for the emptying column 2 is similar in the emptying column 1.

5.4.2 Proposed model definition

In Figure 5.4 is presented a case of two emptying columns. More complex systems can be treated in the same way based on the proposed model.

The corresponding equations of the pipeline are:

1. Mass oscillation equation applied to the emptying column 1

$$\frac{dv_{w,1}}{dt} = \frac{p_1^* - p_{atm}^*}{\rho_w L_{e,1}} + g\frac{\Delta z_{e,1}}{L_{e,1}} - f\frac{v_{w,1}|v_{w,1}|}{2D} - \frac{R_{v,1}gA_1^2 v_{w,1}|v_{w,1}|}{L_{e,1}} \tag{5.13}$$

2. Emptying column 1 position

$$\frac{dL_{e,1}}{dt} = -v_{w,1} \rightarrow L_{e,1} = L_{e,1,0} - \int_0^t v_{w,1}dt \tag{5.14}$$

3. Mass oscillation equation applied to the emptying column 2

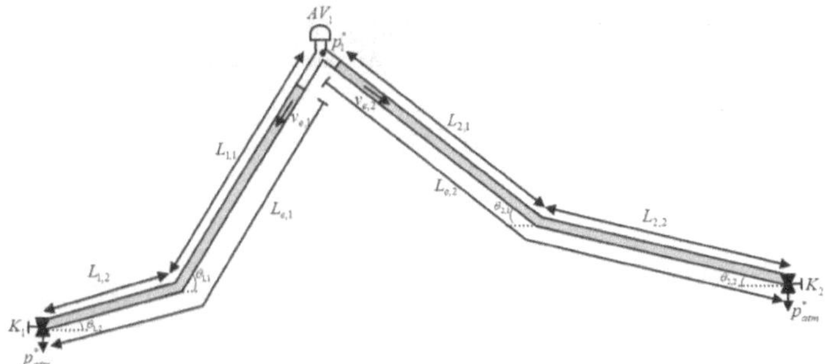

Figure 5.4: Two emptying columns in a pipeline

$$\frac{dv_{w,2}}{dt} = \frac{p_1^* - p_{atm}^*}{\rho_w L_{e,2}} + g\frac{\Delta z_{e,2}}{L_{e,2}} - f\frac{v_{w,2}|v_{w,2}|}{2D} - \frac{R_{v,2}gA_2^2v_{w,2}|v_{w,2}|}{L_{e,2}} \tag{5.15}$$

4. Emptying column 2 position

$$\frac{dL_{e,2}}{dt} = -v_{w,2} \rightarrow L_{e,2} = L_{e,2,0} - \int_0^t v_{w,2}\mathrm{dt} \tag{5.16}$$

5. Evolution of the air pocket

$$\frac{dp_1^*}{dt} = -k\frac{p_1^*(v_{w,1}A_1 + v_{w,2}A_2)}{A_2(L_2 - L_{e,2}) + A_1(L_1 - L_{e,1})} + \frac{p_1^*\rho_{a,nc}v_{a,nc,1}A_{adm,1}}{A_2(L_2 - L_{e,2}) + A_1(L_1 - L_{e,1})}\frac{k}{\rho_{a,1}} \tag{5.17}$$

6. Continuity equation of the air pocket

$$\frac{d\rho_{a,1}}{dt} = \frac{\rho_{a,nc}v_{a,nc,1}A_{adm,1} - (v_{w,1}A_1 + v_{w,2}A_2)\rho_{a,1}}{A_2(L_2 - L_{e,2}) + A_1(L_1 - L_{e,1})} \tag{5.18}$$

7. Air valve characterization

$$Q_{a,nc,1} = C_{adm,1}A_{adm,1}\sqrt{7p_{atm}^*\rho_{a,nc}\left[\left(\frac{p_1^*}{p_{atm}^*}\right)^{1,4286} - \left(\frac{p_1^*}{p_{atm}^*}\right)^{1,714}\right]} \tag{5.19}$$

This set of 7 differential-algebraic equations (5.13 to 5.19) together with the initial condition given by $v_{w,1}(0) = 0$, $L_{e,1}(0) = L_{e,1,0}$, $v_{w,2}(0) = 0$, $L_{e,2}(0) = L_{e,2,0}$, $p_{1,0}^* = p_{atm}^* =$

Table 5.1: Characteristics of tests

Test No.	Air valve model	Air pocket length (m)
1	S050	0,001
2	S050	0,540
3	S050	0,920
4	S050	1,320
5	S050	2,120
6	D040	0,001
7	D040	0,540
8	D040	0,920
9	D040	1,320
10	D040	2,120

101325 Pa, $\rho_{a,1,0} = \rho_{a,nc} = 1,205$ kgm^{-3} and $Q_{a,nc,1}(0) = 0$ allow to describe the whole problem. Simulink Library in Matlab was used in order to compute the seven (7) unknowns functions: $v_{w,1}$, $L_{e,1}$, $v_{w,2}$, $L_{e,2}$, p_1^*, $\rho_{a,1}$ and $Q_{a,nc,1}$.

5.4.3 Experimental results

Ten experimental tests are selected as shown in Table 5.1, where two different air valves and five air pocket sizes were defined. The air valve D040 admits large quantities of air during the emptying process and it has a diameter of 9,375 mm and C_{adm} of 0,375 according with the vacuum curve presented by the manufacturer. The air valve S050 is not recommended for vacuum protection because it has a smaller orifice of 3,175 mm. The manufacturer does not present a vacuum curve because it is used specially for relief in pressurized systems. Consequently, the C_{adm} was calibrated during the tests, with a value of 0,303. The selection of the appropriate air valve size is of utmost importance. The initial air pocket lengths were 0,001 m, 0,540 m, 0,920 m, 1,320 m and 2,120 m. To avoid a numerical problem in the proposed model a minimum air pocket size around of 1 mm is imposed instead of 0 mm.

In the model a constant friction factor of $f = 0,018$ was used. Valves MBV_1 and MBV_2 were modeled by using a synthetic maneuver, with a flow factor of $R_{v,1} = R_{v,2} = 5,1 \bullet 10^5$ m$_{H20}$s^{2}m^{-6} and a valve maneuvering time (T_m) of 1,6 s. The flow factor represents the local losses due to the opening of the valve and the reduction from $DN63$ to $DN25$. The expansion of the air pocket was modeled by using a polytropic model in adiabatic conditions ($k = 1,4$), because the event occurs very fast.

According to the results there are two types of behavior that depend on the air valve: (1) air valve S050 (see Figure 5.5), and (2) air valve D040 (see Figure 5.6). In all tests, the proposed model can predict the subatmospheric pressure pattern. Test No. 1 and Test No. 6 were selected in order to compare results.

Figure 5.5a shows Test No. 1 (air pocket size of 0,001 m) where the absolute pressure head reaches quickly the minimum of subatmospheric pressure of 9,61 m at 1,69 s, then the absolute pressure pattern starts to increase slowly until it reaches the atmospheric condition. The duration of the hydraulic event is 40,3 s. In contrast, when the air valve $D040$ is used then small troughs of subatmospheric pressure occur and the hydraulic event is very short. Figure 5.6a shows the results for Test No. 6 (air pocket size of 0,001 m) where the absolute pressure decreases quickly until it reaches the minimum of subatmospheric pressure head of 10,16 m at 1,82 s and then it increases again until it reaches the atmospheric condition (10,33 m). The duration of the hydraulic event is 8,13 s. Figures 5.5 and 5.6 show that the pressure drop is linear due to the opening of the valves MBV_1 and MBV_2. Then subatmospheric pressure is presented and the water flow starts to decrease since the air valve can admit a better ratio of the air flow, and consequently the pressure pattern rises.

In more complex and large systems, an air valve similar to $S050$ cannot be recommended as protection device during the emptying process because depending on the conditions of the installation, the subatmospheric pressure can reach excessively low values. Engineers should select an air valve similar to $D040$ for minimizing problems associated with the pressure drop to subatmospheric value.

The density of the air pocket is validated with the measurements of the absolute pressure since these variables are related, because the temperature of the air pocket remains practically constant. So, the results are similar considering an isothermal process ($k = 1, 0$).

Figures 5.7 and 5.8 present the evolution of the length of the emptying columns 1 and 2 for all tests. Figure 5.7b shows the results for Test No. 2 (air valve $S050$) where the emptying column 1 reached the horizontal pipe at 28 s, while the emptying column 2 reached it at 29 s. This difference of 1 s was caused because the valves MBV_1 and MBV_2 were not opened exactly at the same time. In contrast, Figure 5.8b shows the results for Test No. 7 (air valve $D040$), where practically the two emptying column reached the horizontal pipe in 5 s because the hydraulic event in this case is faster. In both tests the proposed model predicted the length of emptying columns. It is important to note that when the emptying column reaches the horizontal pipe $\theta = 0°$, the proposed model cannot be applied because air-water interface is parallel to the horizontal pipe direction (as a stratified flow).

Figures 5.9 and 5.10 show the comparison between computed and measured water velocity for all tests. In all of them the water velocity in the emptying column 1 is practically the same as in the emptying column 2 ($v_{w,1} \approx v_{w,2}$). Also, in all tests from 0 s to 1,6 s the water velocity is induced by the opening of the valves MBV_1 and MBV_2. In this range of values, the measurements are not adequate because the system starts resting and the UDV cannot detect the suspended small seeding particles because of no reflection. However, after 1,6 s the proposed model can predict adequately and give information about the system behavior. Figure 5.9c presents the results for the Test No. 3 (air valve $S050$) where rapidly the maximum water velocity is reached at 1,39 s, with a value of 0,076 ms^{-1}. According to the measurements, the maximum value is 0,0775 ms^{-1} at 1,30 s which is very close to the proposed model. After the maximum value attained the water velocity decreases linearly until it reaches a value of 0 ms^{-1} at 34 s. The oscillations around 2 s are caused after the completely opening of the valves MBV_1 and MBV_2. The water velocity range is very low during the emptying procedure with the air valve $S050$, consequently the UDV device with

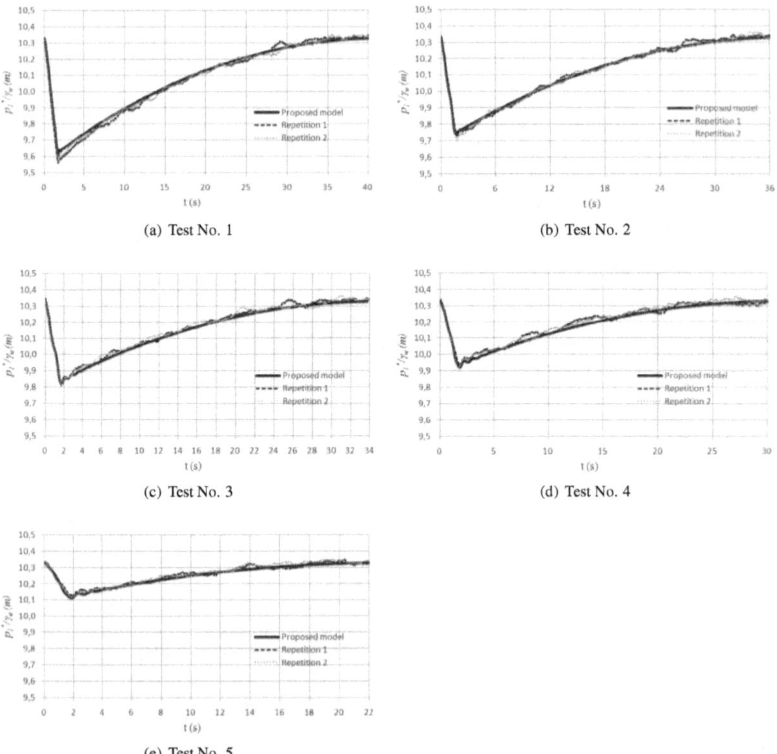

(a) Test No. 1

(b) Test No. 2

(c) Test No. 3

(d) Test No. 4

(e) Test No. 5

Figure 5.5: Comparison between computed and measured absolute pressure oscillation pattern (air valve S050)

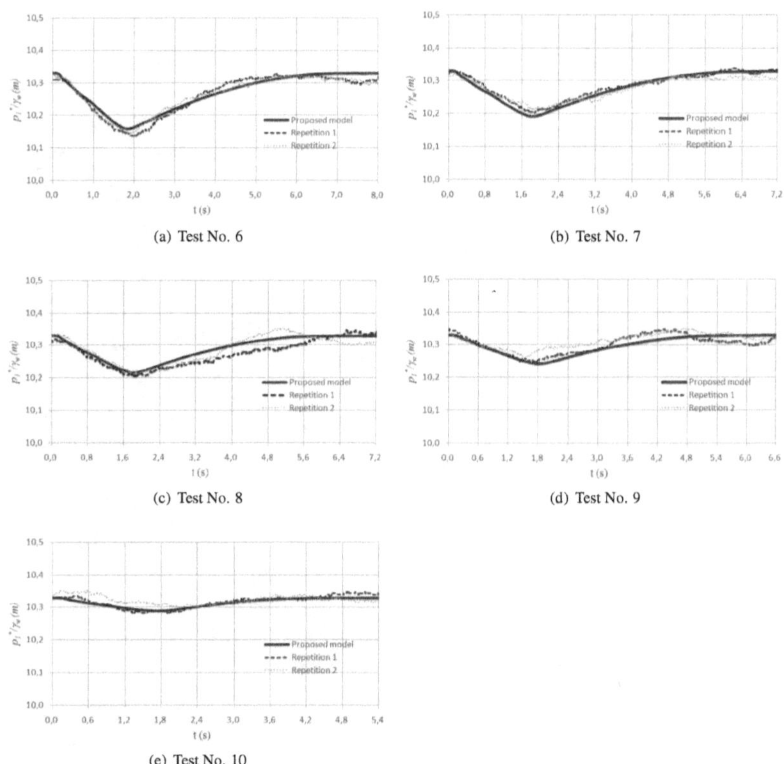

Figure 5.6: Comparison between computed and measured absolute pressure oscillation pattern (air valve D040)

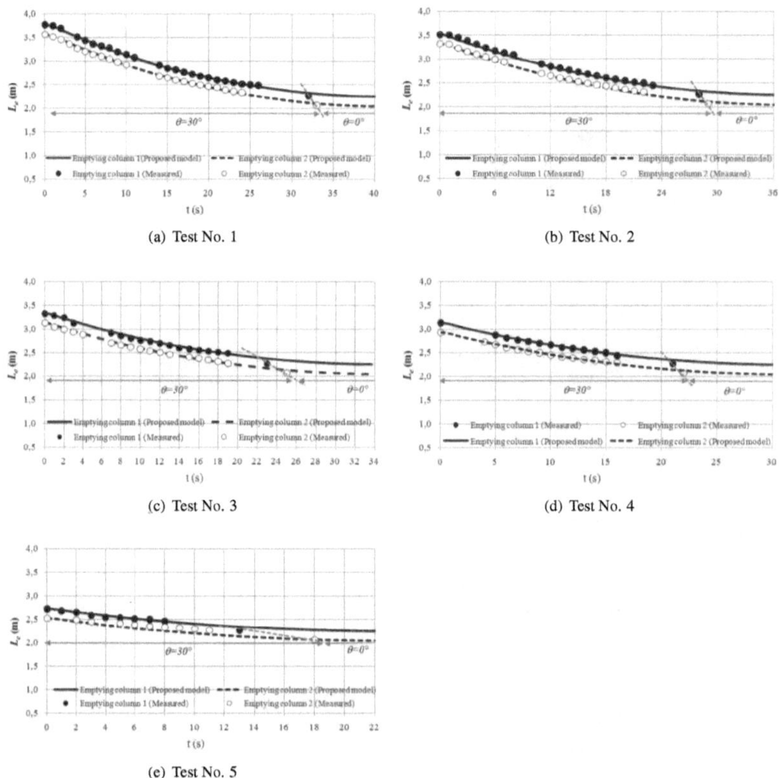

(a) Test No. 1

(b) Test No. 2

(c) Test No. 3

(d) Test No. 4

(e) Test No. 5

Figure 5.7: Comparison between computed and measured length of emptying column (air valve S050)

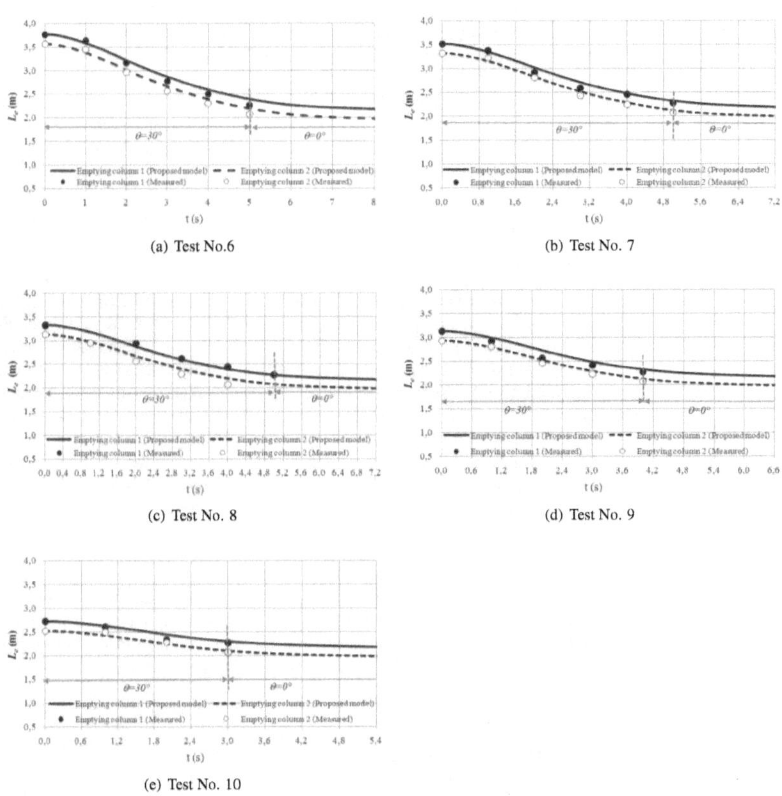

(a) Test No.6

(b) Test No. 7

(c) Test No. 8

(d) Test No. 9

(e) Test No. 10

Figure 5.8: Comparison between computed and measured length of emptying column (air valve D040)

the transducer of $4\ MHz$ of frequency cannot detect appropriately the evolution of the water velocity. It records water velocity with intervals of $0,015$ ms^{-1}. Figure 5.10c shows the comparison between computed and measured water velocities for the Test No. 8 (air valve $D040$) where the water velocity reaches its maximum value of $0,324$ ms^{-1} at $1,77$ s. According to the measurements, the maximum value is $0,32$ ms^{-1} at $1,78$ s which is quite similar to the proposed model. After this maximum the water velocity starts to decrease until the end of the event. During this range, the UDV device can measure appropriately the evolution of the water velocity. The volume of admitted air by $D040$ is almost the water volume drained by valves MBV_1 and MBV_2 since the minimum of subatmospheric pressure head is $10,22$ m, practically the atmospheric pressure.

5.4.4   Sensitivity analysis

Effect of air pocket sizes

It is important to identify the great influence of the size of the entrapped air pocket on the minimum of the subatmospheric pressure. Figure 5.11 shows the results taking different air pocket lengths ($0,001$ m, $0,540$ m, $0,920$ m, $1,320$ m, and $2,120$ m). The smallest the air pocket size is, the highest trough of subatmospheric pressure is noticed. Equation 5.17 shows this situation with the comparison between the gradient of the absolute pressure (dp_i^*/dt) and the air pocket volume ($A_2(L_2 - L_{e,2}) + A_1(L_1 - L_{e,1})$). For the air valve $S050$, subatmospheric pressure head are found in the range of $9,61$ m and $10,1$ m, while for the air valve $D040$ small variations are found between $10,16$ m and $10,32$ m showing the adequacy of this air valve for the emptying process. It also shows the importance of the air valve size since it can induce critical conditions associated with the subatmospheric pressure occurrence.

Maximum water velocity

Figure 5.12 shows a comparison between computed and measured maximum water velocities. The proposed model predicts the maximum values of the water velocity for all tests. For the air valve $S050$, maximum values of water velocity are found in the range of $0,049$ ms^{-1} to $0,079$ ms^{-1}, while for the air valve $D040$ water velocities are found in the range of $0,193$ ms^{-1} to $0,397$ ms^{-1}.

5.5   Case study and results

5.5.1   Description of case study

The proposed model is used to simulate the emptying process of a pipeline of irregular profile with four branches and three air pockets. The system has the following data: $D = 0,3$ m, $L_1 = 700$ m (with length branches $L_{1,1}$ and $L_{1,2}$, and slope branches $\theta_{1,1}$ and $\theta_{1,2}$), $L_2 = 600$ m (with slope branch θ_2), $L_3 = 600$ m (with slope branch θ_3), and $L_4 = 700$ m (with length branches $L_{4,1}$, $L_{4,2}$, and $L_{4,3}$, and slope branches $\theta_{4,1}$, $-\theta_{4,2}$, and $\theta_{4,3}$). The

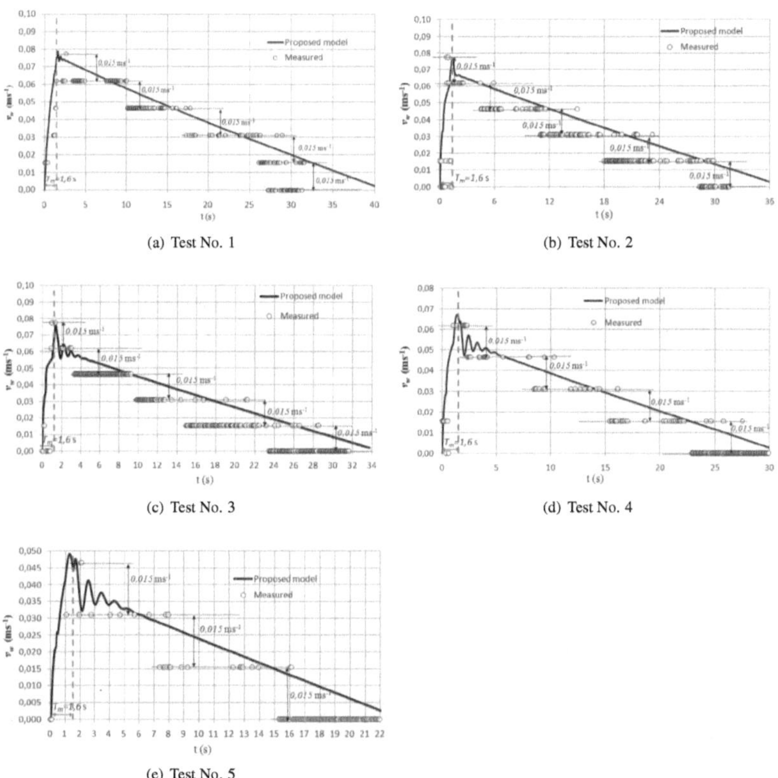

(a) Test No. 1

(b) Test No. 2

(c) Test No. 3

(d) Test No. 4

(e) Test No. 5

Figure 5.9: Comparison of water velocity between computed and measured (air valve $S050$)

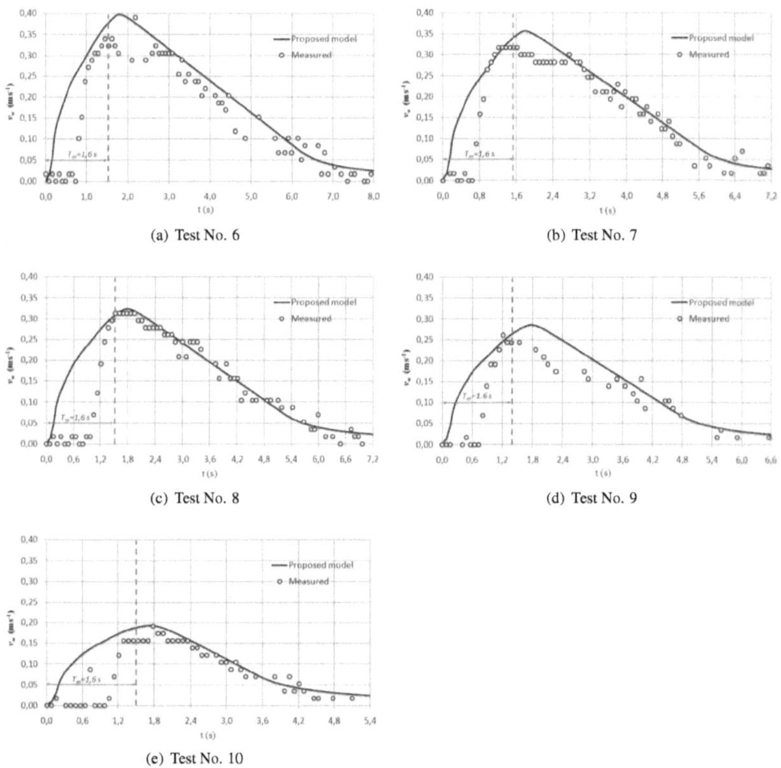

(a) Test No. 6

(b) Test No. 7

(c) Test No. 8

(d) Test No. 9

(e) Test No. 10

Figure 5.10: Comparison of water velocity between computed and measured (air valve D040)

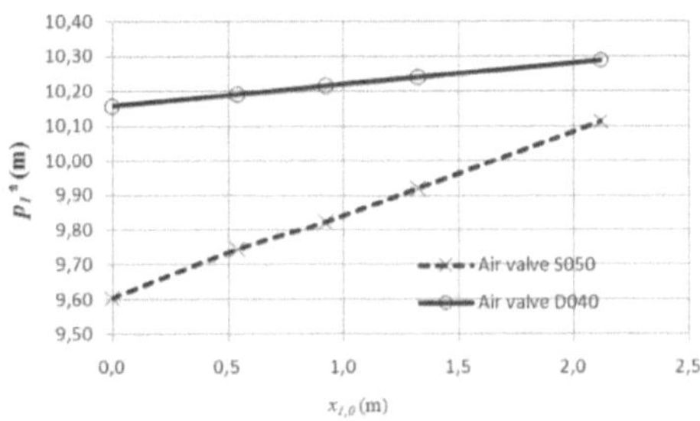

Figure 5.11: Effect of the air pocket sizes on the minimum pressure attained

drainage of the system is carried out by drain valves DV_1 and DV_2 with a flow resistance coefficient of $K_1 = K_2 = 3{,}02$ m^3/s/m$^{1/2}$. Air valves 1 and 3 have a nominal diameter of 250 mm, and air valve 2 of 350 mm. All air valves have an inflow discharge coefficient of 0,50. The air pocket sizes are x_1, x_2, and x_3 with absolute pressure values of p_1^*, p_2^*, and p_3^*, respectively. Figure 5.13 shows characteristics of the installation.

5.5.2 Proposed model definition

The corresponding equations of the pipeline are:

1. Mass oscillation equation applied to the emptying column 1

$$\frac{dv_1}{dt} = \frac{p_1^* - p_{atm}^*}{\rho_w L_{w,1}} + g\frac{\Delta z_1}{L_{w,1}} - f\frac{v_1|v_1|}{2D} - \frac{gA^2(v_1 + v_2)|v_1 + v_2|}{K_1^2 L_{w,1}} \tag{5.20}$$

2. Air-water interface for the water column 1

$$\frac{dL_{w,1}}{dt} = -v_1 \rightarrow L_{w,1} = L_{w,1,0} - \int_0^t v_1 dt \tag{5.21}$$

3. Mass oscillation equation applied to the emptying column 2

$$\frac{dv_2}{dt} = \frac{p_2^* - p_{atm}^*}{\rho_w L_{w,2}} + g\frac{\Delta z_2}{L_{w,2}} - f\frac{v_2|v_2|}{2D} - \frac{gA^2(v_1 + v_2)|v_1 + v_2|}{K_1^2 L_{w,2}} \tag{5.22}$$

4. Air-water interface for the water column 2

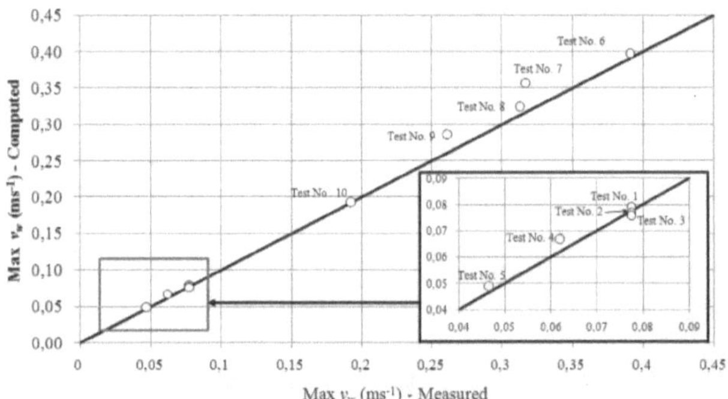

Figure 5.12: Comparison between computed and measured maximum water velocities

$$\frac{dL_{w,2}}{dt} = -v_2 \rightarrow L_{w,2} = L_{w,2,0} - \int_0^t v_2 dt \tag{5.23}$$

5. Mass oscillation equation applied to the emptying column 3

$$\frac{dv_3}{dt} = \frac{p_3^* - p_{atm}^*}{\rho_w L_{w,3}} + g\frac{\Delta z_3}{L_{w,3}} - f\frac{v_3|v_3|}{2D} - \frac{gA^2(v_3 + v_4)|v_3 + v_4|}{K_2^2 L_{w,3}} \tag{5.24}$$

6. Air-water interface for the water column 3

$$\frac{dL_{w,3}}{dt} = -v_3 \rightarrow L_{w,3} = L_{w,3,0} - \int_0^t v_3 dt \tag{5.25}$$

7. Mass oscillation equation applied to the emptying column 4

$$\frac{dv_4}{dt} = \frac{p_3^* - p_{atm}^*}{\rho_w L_{w,4}} + g\frac{\Delta z_4}{L_{w,4}} - f\frac{v_4|v_4|}{2D} - \frac{gA^2(v_3 + v_4)|v_3 + v_4|}{K_2^2 L_{w,4}} \tag{5.26}$$

8. Air-water interface for the water column 4

$$\frac{dL_{w,4}}{dt} = -v_4 \rightarrow L_{w,4} = L_{w,4,0} - \int_0^t v_4 dt \tag{5.27}$$

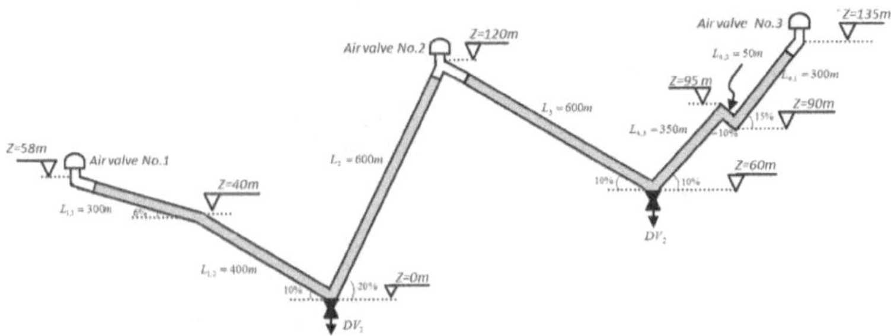

Figure 5.13: Pipeline of irregular profile with air valves

9. Evolution of the air pocket 1

$$\frac{dp_1^*}{dt} = -k\frac{p_1^*}{L_1 - L_{w,1}}v_1 + \frac{p_1^*}{A(L_1 - L_{w,1})}\frac{k}{\rho_{a,1}}\rho_{a,nc}v_{a,nc,1}A_{adm,1} \qquad (5.28)$$

10. Continuity equation of the air pocket 1

$$\frac{d\rho_{a,1}}{dt} = \frac{\rho_{a,nc}v_{a,nc,1}A_{adm,1} - v_1 A\rho_{a,1}}{A(L_1 - L_{w,1})} \qquad (5.29)$$

11. Air valve 1 characterization

$$Q_{a,nc,1} = C_{adm,1}A_{adm,1}\sqrt{7p_{atm}^*\rho_{a,nc}\left[\left(\frac{p_1^*}{p_{atm}^*}\right)^{1,4286} - \left(\frac{p_1^*}{p_{atm}^*}\right)^{1,714}\right]} \qquad (5.30)$$

12. Evolution of the air pocket 2

$$\frac{dp_2^*}{dt} = -k\frac{p_2^*(v_2 + v_3)}{L_2 - L_{w,2} + L_3 - L_{w,3}} + \frac{p_2^*}{A(L_2 - L_{w,2} + L_3 - L_{w,3})}\frac{k}{\rho_{a,2}}\rho_{a,nc}v_{a,nc,2}A_{adm,2} \qquad (5.31)$$

13. Continuity equation of the air pocket 2

$$\frac{d\rho_{a,2}}{dt} = \frac{\rho_{a,nc}v_{a,nc,2}A_{adm,2} - (v_2 + v_3)A\rho_{a,2}}{A(L_2 - L_{w,2} + L_3 - L_{w,3})} \qquad (5.32)$$

14. Air valve 2 characterization

$$Q_{a,nc,2} = C_{adm,2} A_{adm,2} \sqrt{7 p_{atm}^* \rho_{a,nc} \left[\left(\frac{p_2^*}{p_{atm}^*} \right)^{1,4286} - \left(\frac{p_2^*}{p_{atm}^*} \right)^{1,714} \right]} \qquad (5.33)$$

15. Evolution of the air pocket 3

$$\frac{dp_3^*}{dt} = -k \frac{p_3^*}{L_4 - L_{w,4}} v_4 + \frac{p_3^*}{A(L_4 - L_{w,4})} \frac{k}{\rho_{a,3}} \rho_{a,nc} v_{a,nc,3} A_{adm,3} \qquad (5.34)$$

16. Continuity equation of the air pocket 3

$$\frac{d\rho_{a,3}}{dt} = \frac{\rho_{a,nc} v_{a,nc,3} A_{adm,3} - v_4 A \rho_{a,3}}{A(L_4 - L_{w,4})} \qquad (5.35)$$

17. Air valve 3 characterization

$$Q_{a,nc,3} = C_{adm,3} A_{adm,3} \sqrt{7 p_{atm}^* \rho_{a,nc} \left[\left(\frac{p_3^*}{p_{atm}^*} \right)^{1,4286} - \left(\frac{p_3^*}{p_{atm}^*} \right)^{1,714} \right]} \qquad (5.36)$$

A 17x17 system of DAE describe the whole problem. Together with the corresponding boundary and initial conditions, 17 unknowns can be solved v_1, $L_{w,1}$, v_2, $L_{w,2}$, v_3, $L_{w,3}$, v_4, $L_{w,4}$, p_1^*, p_2^*, p_3^*, $\rho_{a,1}$, $\rho_{a,2}$, $\rho_{a,3}$, $v_{a,cn,1}$, $v_{a,cn,2}$, $v_{a,cn,3}$. The initial conditions at $t = 0$ are described by $v_1(0) = 0$, $L_{w,1}(0) = L_{w,1,0}$, $v_2(0) = 0$, $L_{w,2}(0) = L_{w,2,0}$, $v_3(0) = 0$, $L_{w,3}(0) = L_{w,3,0}$, $v_4(0) = 0$, $L_{w,4}(0) = L_{w,4,0}$, $p_1^*(0) = p_{atm}^*$, $p_2^*(0) = p_{atm}^*$, $p_3^*(0) = p_{atm}^*$, $\rho_{a,1,0} = 1{,}205$ kg/m^3, $\rho_{a,2,0} = 1{,}205$ kg/m^3, $\rho_{a,3,0} = 1{,}205$ kg/m^3, $v_{a,cn,1}(0) = 0$, $v_{a,cn,2}(0) = 0$, and $v_{a,cn,3}(0) = 0$. The upstream boundary conditions are given by $p_{1,0}^*$, $p_{2,0}^*$, and $p_{3,0}^*$ (initial conditions of air pockets), and the downstream boundary condition are given by p_{atm}^* (water free discharge to the atmosphere). Gravity terms $\Delta z_1 / L_{w,1}$, $\Delta z_2 / L_{w,2}$, $\Delta z_3 / L_{w,3}$, and $\Delta z_4 / L_{w,4}$ can be computed as presented by Coronado-Hernández et al. (2017b).

5.5.3 Results

Figure 5.14 shows results for the pipeline of irregular profile with air valves for the main hydraulic and thermodynamic variables along the transient event (water flow, length of the water column, absolute pressure, air density, and air flow). The initial air pocket sizes were $x_{1,0} = x_{2,0} = x_{3,0} = 1$ m, considering the most critical situation (Coronado-Hernández et al., 2017b). A polytropic coefficient (k) of 1,2, and a friction factor (f) of 0,018 were considered.

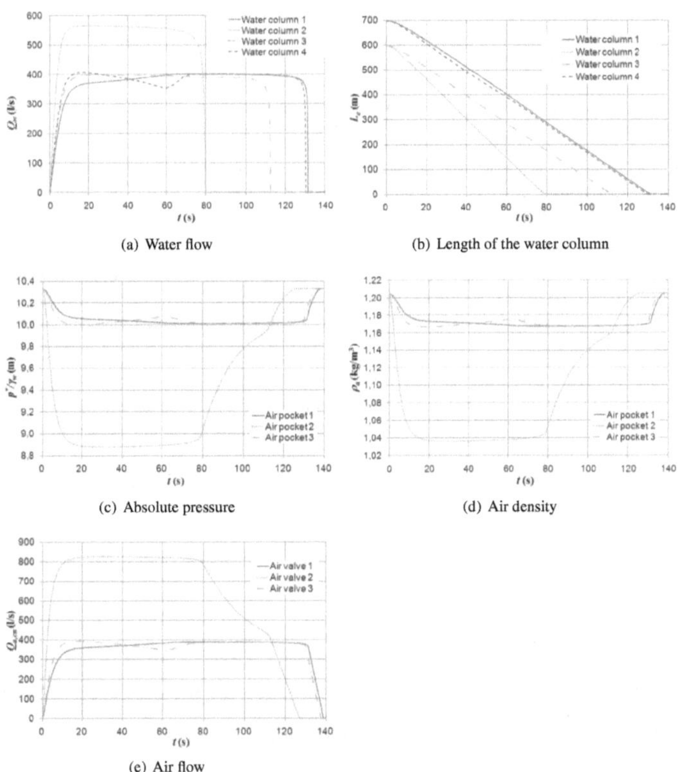

(a) Water flow

(b) Length of the water column

(c) Absolute pressure

(d) Air density

(e) Air flow

Figure 5.14: Full transient analysis of the pipeline of irregular profile with air valves

According to Figure 5.14, the authors can deduce:

- In water column 1, the maximum water flow during the transient occurs at 70,2 s, with a value of 401,9 ls^{-1} (see Figure 5.14a). Two types of behavior are presented because there are two branches. Water column 1 and 2 present a similar behavior with maximum values of 565,3 ls^{-1} and 397,7 ls^{-1}, respectively. Regarding to the water column 4, the negative slope branch of -10% produces an effect of the water flow pattern as shown from 57,8 s to 67,5 s. The maximum value is reached at 16,6 s (406,5 ls^{-1}).

- Figure 5.14b shows that the water columns are drained completely indicating that air valves are well sized. The water column behaviors are linear. The drainage times for water column 1 trough 4 are in range from 79,5 s to 131,3 s.

- Air pockets are at atmospheric pressure (101325 Pa). Air pockets 1 and 3 present a similar behavior because they are admitting air for water columns 1 and 4. The minimum absolute pressure head is 10,0 m. Air pocket 2 has the lowest values of subatmospheric pressure head with 8,9 m because air valve 2 has to admit simultaneously air into water column 2 and 3, as shown in Figure 5.14c.

- The change in absolute pressure head parallels the air density in the air pockets (see Figure 5.14d). The minimum values for air pockets 1, 2, and 3 are 1,17 kgm^{-3}, 1,04 kgm^{-3}, and 1,17 kgm^{-3}, respectively. Air pockets 1 and 3 present a similar behavior because are located at the ends of the pipeline.

- Figure 5.14e presents the behavior of admitted air. Air valves 1 and 3 admit lower air volume than air valve 2 because they are entering air into a one pipe. The maximum air flows are reached for air valve 1 at 77,1 s with a value of 389,2 ls^{-1}; and for air valve 3 at 17,5 s (393,5 ls^{-1}). Air valve 2 has to admit air for two pipes. The maximum value is 827,0 ls^{-1} (at 30,3 s).

5.6 Conclusions

Subatmospheric pressures occur during the emptying process in pipelines with undulating profile, which is a typical and common operation that engineers have to face. A rigid two-phases flow model was developed for analyzing it considering n possible air pockets, d air valves, p pipes, and o drain valves. The proposed model is validated using an experimental facility of irregular profile of 7,3 m long and nominal diameter 63 mm ($DN63$) with an air valve located at the high point and several drain valves along the pipeline. The main hydraulic variables could be measured such as flow (both phases), pressure and air-water front position. Comparisons between computed and measured values of the absolute pressure, water velocity and the length of the emptying columns show that the proposed model can predict accurately not only the extreme values but also the patterns of them. According to the results, the following conclusions can be drawn:

1. The proposed model can be used for planning the emptying operations in pipelines with undulating profile considering its limitations.

2. The expansion of the air pocket produces minimum subatmospheric pressure. In order to control these effects, air valves should be installed in pipelines.

3. It is very important to have a correct selection of air valves for vacuum protection for emptying pipelines. An inadequate selection produces both lower values of subatmospheric pressure and a slower drainage of the system. On the other hand, a larger air valve orifice size reduces the lowest values of subatmospheric pressure.

4. Engineers should consider for the initial condition the pipeline is completely filled, which is the most critical condition, since the smallest air pocket sizes produce the lowest subatmospheric pressures.

In real hydraulic pressurized systems the proposed model can be used for: (i) checking the risk of a system collapse considering factors such as stiffness class, the soil in natural conditions, the type of backfill and the cover depth, and (ii) selecting the air valve sizes depending on the characteristics of each hydraulic system.

Notation

A	Cross sectional area of the pipe (m^2)
$A_{adm,m}$	Cross section area of the air valve m (m^2)
$C_{adm,m}$	Inflow discharge coefficient of the air valve m $(-)$
D	Internal pipe diameter (m)
f	Darcy-Weisbach friction factor $(-)$
g	Gravity acceleration (m/s^2)
$L_{e,j}$	Length of the emptying column j (m)
L_j	Total length of the pipe j (m)
k	Polytropic coefficient $(-)$
K_s	Flow factor of the drain valve s (m^3/s) with a pressure drop of 1 m_{H20}
$h_{m,s}$	Local loss of the drain valve s (m)
$m_{a,i}$	Air mass of the air pocket i (kg)
p_i^*	Absolute pressure of the air pocket i (Pa)
p_{atm}^*	Atmospheric pressure (Pa)
t	Time (s)
T_m	Valve maneuvering time (s)
$Q_{a,nc,m}$	Air discharge in normal conditions admitted by the air valve m (m^3/s)
$Q_{w,s}$	Water discharge by the drain valve s (m^3/s)
$V_{a,i}$	Air volume of the air pocket i (m^3)
$v_{e,j}$	Water velocity of the emptying column j (m/s)
$v_{a,nc,m}$	Air velocity in normal conditions admitted by the air valve m (m/s)
x_i	Length of the air pocket i (m)
$\Delta z_{e,j}$	Elevation difference (m)
$\rho_{a,i}$	Air density of the air pocket i (kg/m^3)
$\rho_{a,nc}$	Air density in normal conditions (kg/m^3)
ρ_a	Water density (kg/m^3)

Superscripts

$*$	Absolute values

Subscripts

a	Refers to air
i	Refers to air pocket
j	Refers to pipe
m	Refers to air valve
nc	Normal conditions
w	Refers to water
0	Initial condition

Chapter 6

Rigid water column model for simulating the emptying process in a pipeline using pressurized air

This chapter is extracted from the paper:

Rigid water column model for simulating the emptying process in a pipeline using pressurized air
Coauthors: Coronado-Hernández O. E; Fuertes-Miquel, V. S; Iglesias-Rey, P.L.; Martínez-Solano, F.J.
Journal: Journal of Hydraulic Engineering ISSN 0733-9429.
2017 Impact Factor: 2.080. Position JCR 39/128 (Q2). Civil Engineering.
State: Published. 2018. Volume 144; Issue 4. DOI: 10.1061/(ASCE)HY.1943-7900.0001446.

6.2   Introduction

Studies related to the behavior of entrapped air during filling and emptying maneuvers in water supply networks are very complex to analyze because there are two fluids (water and air) that exist in two different phases (liquid and gas) (Fuertes, 2001).

Filling and emptying processes can be analyzed using inertial models: the elastic water model (EWM) and the rigid water column model (RWCM). The EWM considers the elasticity of water and the pipe, while the RWCM ignores them. The solution is obtained using numerical methods (Zhou et al., 2011b).

Liou and Hunt (1996) developed an RWCM for analyzing the filling process that only considers the evolution of the water column. Izquierdo et al. (1999) developed an RWCM that considers not only the evolution of the water column but also the air-water interface and the gravity term to represent the irregular profile in the pipeline. Zhou et al. (2013b) developed an EWM to analyze a rapidly filling process.

The emptying process is the reverse process of the filling process in pipelines; however, the emptying process in pipelines has not yet been studied as comprehensively. These studies are very important because pipelines must be emptied periodically. Laanearu et al. (2012) carried out experiments on a PVC-steel pipeline with a 232 mm internal diameter and a 271, 6 m length with the primary objective of providing data for future validation. This installation setup consisted of a constant-head water supply tower, a high-pressure air tank, a 261 m long horizontal PVC pipe, a 10,6 m long steel pipe (divided into a 6,1 m long horizontal pipe and a 4,5 m long vertical pipe), PVC and steel joints, steel inlet and outlet parts, various types of valves along the PVC-steel pipeline and a free-surface basement reservoir. Tijsseling et al. (2016) developed an RWCM for analyzing the emptying process using pressurized air. That model assumed that the velocity of the water column was not uniform because the movement of the front and tail ends were different.

In this paper, a mathematical model for the emptying process in a pipeline using pressurized air is developed based on two physical equations. The first equation is the mass oscillation equation, which provides sufficient accuracy for modeling this phenomenon (Cabrera et al., 1992; Liou and Hunt, 1996; Izquierdo et al., 1999). The second equation is the air-water interface equation (Zhou et al., 2013a,b). The gravity term is added in this model to represent the irregular profile of the pipeline. For validation, the authors use the data provided by Laanearu et al. (2012). Comparisons between both the computed and measured water flow oscillations and gauge pressures are conducted along the pipeline. In addition, comparisons between the proposed model and previous models are conducted. Finally, a sensitivity analysis of the height of the vertical pipe is performed.

6.3   Mathematical model

Fig. 6.1 shows the scheme of the pipeline, which consists of a high-pressure air tank, a length of the emptying column, a regulating valve located at the downstream end, a horizontal pipe and a vertical pipe at the end.

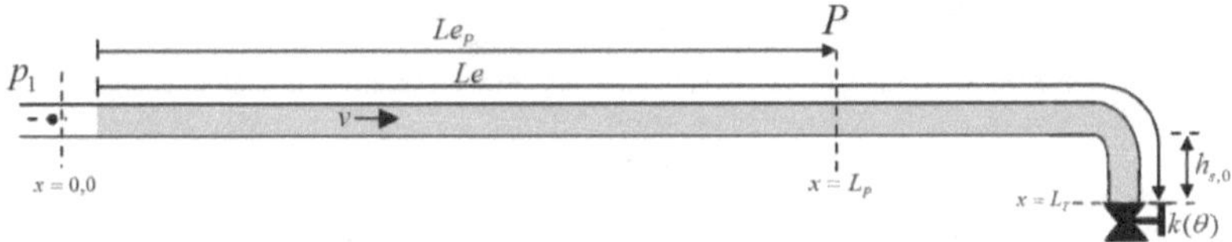

Figure 6.1: Schematic of pressurized air in a water-emptying horizontal-vertical pipeline

6.3.1 Equations

The proposed model considers uniform movement within the water column. The following assumptions are made:

- The water behavior is modeled using the rigid model approach.

- The air-water interface has a well-defined cross section.

- Constant friction accounts for the friction losses.

Based on these assumptions, this problem is modeled using the following equations:

- The mass oscillation equation for an emptying column (rigid water column approach):

$$\frac{dv}{dt} = \frac{p_1}{\rho_w Le} + g\frac{h_s}{Le} - f\frac{v|v|}{2D} - k(\theta)\frac{v|v|}{2Le} \qquad (6.1)$$

- The interface position of the emptying column:

$$\frac{dLe}{dt} = -v \quad \left(Le = Le_0 - \int_0^t v\,dt\right) \qquad (6.2)$$

where p_1 = driving gauge pressure , Le = length of the emptying column at time t, h_s = length of the vertical pipe, v = water velocity, D = internal pipe diameter, $k(\theta)$ = minor loss coefficient of the valve, x = axial coordinate, L_T = pipe length, f = pipe wall friction coefficient, g = gravity acceleration, ρ_w = water density, Le_0 = initial length of emptying column at $t = 0$ and h_s = length of the emptying column at vertical steel pipe.

In summary, a 2x2 system of differential equations (Eq. 6.1 and Eq. 6.2) describes the whole system. Together with the corresponding boundary and initial conditions, the system of equations can be solved for the 2 unknowns: v and Le.

This process involves the development of a complex model. The solution was calculated using Simulink in MatLab.

6.3.2   Initial and boundary conditions

The system is assumed to be initially static at $t = 0$. Therefore, the initial conditions are
described by $v(0) = 0$, $Le(0) = Le_0$ and $h_s(0) = h_{s,0}$.
   The boundary conditions are as follows:

a. The upstream boundary condition is the manometric pressure (p_1) given by the air tank.

b. The downstream boundary condition is the valve loss described by $k(\theta)$. Water is free
to discharge to the atmosphere at $p_{atm} = 0$.

6.3.3   Gravity term

The gravity term (h_s/Le) in Eq. 1 depends on the position of the emptying column. For this
case, there are two possibilities:

a. When the air-water front of the emptying column has not reached the vertical pipe
($Le >= h_{s,0}$):

$$\frac{h_s}{Le} = \frac{h_{s,0}}{Le} \tag{6.3}$$

b. When the air-water front of the emptying column has reached the vertical pipe ($Le <$
$h_{s,0}$):

$$\frac{h_s}{Le} = 1 \tag{6.4}$$

If the configuration setup is different, then the gravity term will be different.

6.3.4   Pressure inside of pipeline

The manometric pressure along the pipeline (at point P) can be computed as follows:

$$p_P = p_1 - \rho_w Le_P \left(\frac{f}{2D} v|v| + \frac{dv}{dt} \right) \tag{6.5}$$

where Le_P = length of the emptying column until point P at time t.
To calculate Le_P, the following expressions can be used:

$$Le_P = \begin{cases} L_P - L_T + Le, & L_T - Le < Le_P \\ 0, & L_T - Le \geq Le_P \end{cases} \tag{6.6}$$

6.4 Numerical model validation

The proposed model was applied to conditions based on experiments conducted by Laanearu et al. (2012). Fig. 6.2 presents a scheme of this installation. For this analysis, an average internal diameter of 232 mm was used, as suggested by Tijsseling et al. (2016).

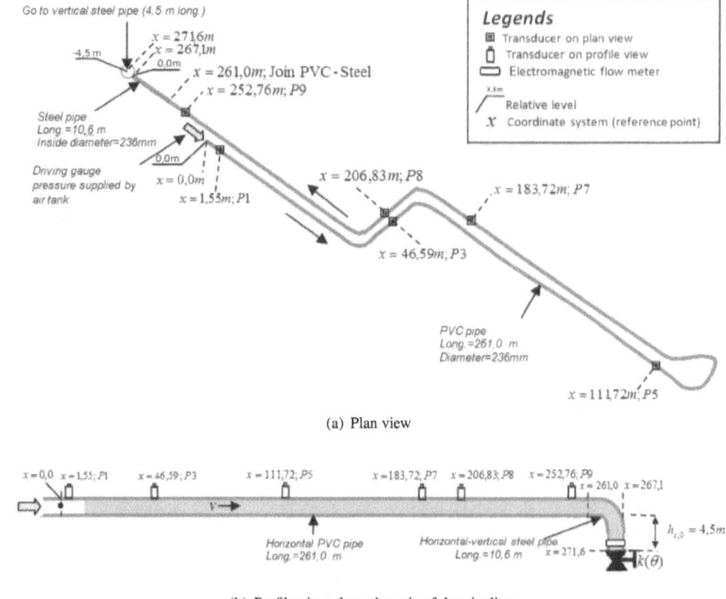

(a) Plan view

(b) Profile view along the axis of the pipeline

Figure 6.2: Schematic of the PVC-steel pipeline and instrumentation

The upstream boundary conditions for the 9 runs were provided by the driving air-pressure head values that were based on measurements made by Laanearu et al. (2012), as shown in Fig. 6.3.

The downstream boundary conditions were provided by both the water freely discharged to the atmosphere ($p_{atm} = 0$) and the position of the manual butterfly valve. According to the experiments of Laanearu et al. (2012), the minor-loss coefficients $k(\theta)$ for the 9 runs were

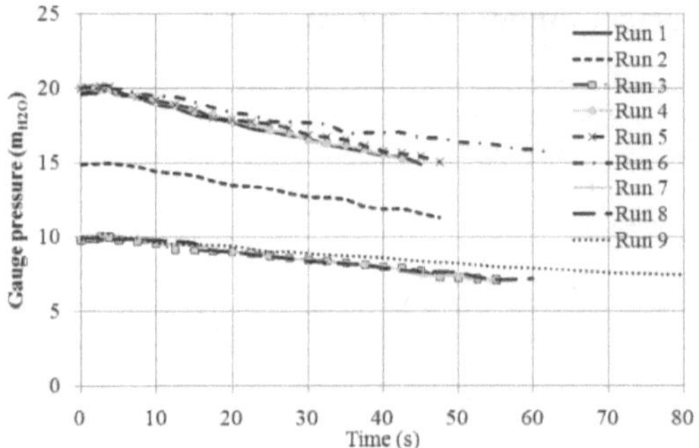

Figure 6.3: Gauge pressure supplied by the air tank during the transient stage. Data are based
on experiments conducted by Laanearu et al. (2012).

3,32, 3,50, 3,48, 3,64, 5,88, 21,24, 3,84, 6,14 and 22,68.

6.4.1   Proposed model verification

To validate the model, comparisons between both the computed and measured water flow
oscillation patterns and gauge pressure patterns were conducted. A friction factor of $f =
0,0117$ was used.

The water flow was determined at the downstream end for the 9 runs by varying the opening of the manual butterfly valve. Fig. 6.4 shows a comparison of the computed and measured
water flow oscillation patterns for runs 1, 4, 5 and 9 at the installation. The comparisons indicate that the water flow oscillations from the model are similar to those of the experiments.
Consequently, the proposed model can effectively simulate the transient flow during an emptying process using pressurized air in a pipeline. For all runs, few results were affected by
varying the opening of the manual butterfly valve.

Table 6.1 compares all 9 runs. For each run, the initial length of the emptying column
and the location of the air-water front were considered. Comparisons were made between
sections 1 and 9 located at $x = 1,55$ m and $x = 252,76$ m, respectively. The proposed model
shows good overall agreement with the experimental data for water velocities at section 1
(v_1) and section 9 (v_9). Greater differences are presented in section 1 for runs 2 and 3 and in
section 9 for run 1. The model presents lower values of τ_{1-9} at these conditions compared to
the measured data.

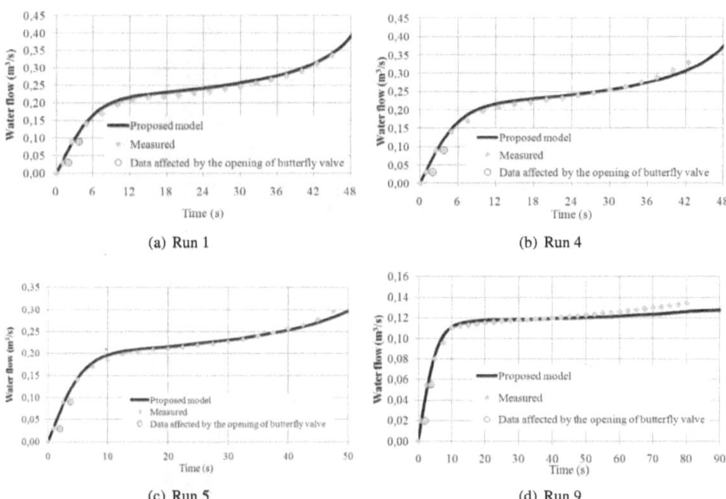

Figure 6.4: Comparisons between the calculated and measured water flow oscillation patterns for runs 1, 4, 5 and 9

Table 6.1: Summary of the emptying process for the 9 runs

| Run No. | $k(\theta)$ [a] | $x_{i,0}^{a}$ | Le_0^{b} | Meas.[c] | Calc.[d] | Meas.[c] | Calc.[d] | Meas.[c] | Calc.[d] |
	(-)	(m)	(m)	v_1 (ms^{-1})	v_1 (ms^{-1})	v_9 (ms^{-1})	v_9 (ms^{-1})	τ_{1-9} (s)	τ_{1-9} (s)
1	3,32	−16,2	287,8	4,18	4,28	7,90	9,13	37	40
2	3,50	−20,8	292,4	2,83	4,03	7,04	7,98	45	46
3	3,48	−20,7	292,3	2,29	3,46	6,20	6,85	51	53
4	3,64	−14,7	286,3	4,20	4,17	8,13	8,84	36	41
5	5,88	−14,7	286,3	4,07	4,07	6,91	7,34	40	45
6	21,24	−13,1	284,7	3,43	3,39	4,16	4,16	54	65
7	3,84	−12,9	284,5	3,20	3,11	6,38	6,59	46	54
8	6,14	−13,8	285,4	3,09	3,07	5,42	5,53	50	59
9	22,68	−15,8	287,4	2,55	2,63	3,09	3,04	69	83

Notes:

[a] refers to the data measured by Laanearu et al. (2012). [b] refers to the data computed at $x_{i,0} + 271,6$ m. [c] refers to the measured data. [d] refers to the data computed by the current model. $x_{i,0}$ is the initial air-water front coordinate, Le_0 is the initial length of the emptying column, v_1 and v_9 are the outflow velocities when the air-water front passes by sections 1 and 9, respectively, and τ_{1-9} is the time between sections 1 and 9.

For run 4, the evolution of the gauge pressure along the PVC-steel pipeline at locations P3 ($x = 46{,}59$ m), P5 ($x = 111{,}72$ m), P7 ($x = 183{,}72$ m) and P8 ($x = 206{,}83$ m) was determined. The proposed model can also approximately reproduce the experimental gauge pressure along the PVC-steel pipeline during the transient phenomenon for run 4 (see Fig. 6.5). At locations P3, P5, P7 and P8, the model excels at predicting the water flow along the PVC-steel pipeline. In all measurements, during the first 15 s, transient oscillations were observed by the gauge pressure readings due to opening the manual butterfly valve. To calculate these oscillations, the opening curve of the ball valve is required. At locations P3 and P5, the proposed model can properly reproduce the measurements after the oscillations have finished. At locations P7 and P8, the model can predict the maximum value of the gauge pressure after 1,7 s and 2,2 s, respectively. The proposed model can successfully predict the peak pressure, which is very important for ensuring the safety of the pipeline.

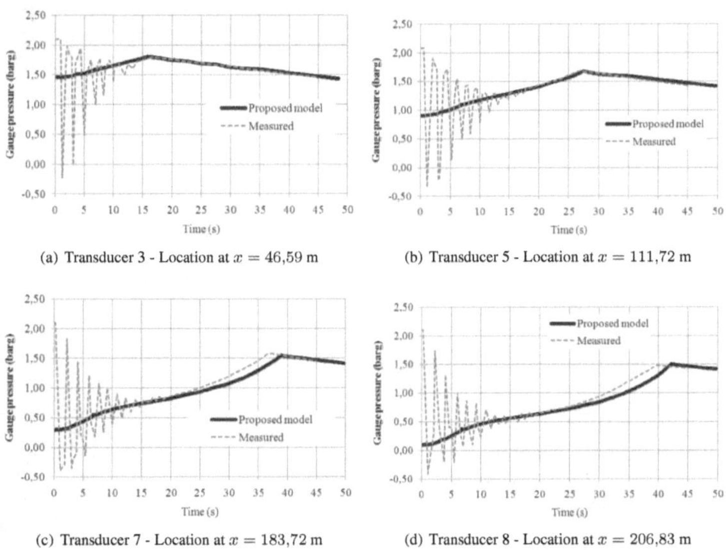

(a) Transducer 3 - Location at $x = 46{,}59$ m

(b) Transducer 5 - Location at $x = 111{,}72$ m

(c) Transducer 7 - Location at $x = 183{,}72$ m

(d) Transducer 8 - Location at $x = 206{,}83$ m

Figure 6.5: Comparisons between the calculated and measured gauge pressure oscillation patterns for run 4

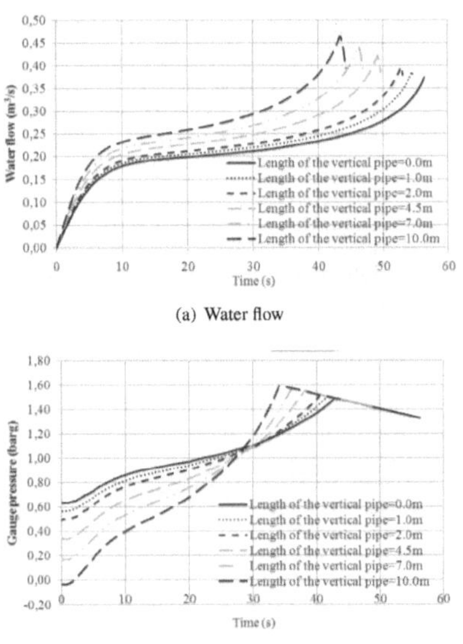

(a) Water flow

(b) Gauge pressure at Transducer 7, located at $x = 183,72$ m

Figure 6.6: Effects of the length of the vertical steel pipe

6.4.2 Influence of the length of the steel pipe

It is important to consider the considerable influence of the length of the steel vertical pipe ($h_{s,0}$) on the emptying process; therefore, values of 1,0 m, 2,0 m, 4,5 m, 7,0 m and 10,0 m were considered to determine the water flow and the gauge pressure variations (see Fig. 8.8). The pipe length (L_T) remained constant in all cases.

Fig. 8.8a presents the influence of the vertical steel pipe length on the water flow. The longer the length of the vertical pipe ($h_{s,0}$) is, the faster the draining process occurs because the gravity term is increased. In this case, the emptying times range between 56,3 s and 44,3 s. In addition, the higher the length of vertical pipe is, the higher the water flows during the transient stage are; the maximum values are between 0,37 m^3s^{-1} and 0,46 m^3s^{-1}. At the end of the transient stage, the water flow has a rapidly decreasing linear profile inside the vertical pipe. When $h_{s,0} = 0$, the installation does not have this tendency.

Fig. 8.8b shows the influence of the length of the vertical steel pipe on the gauge pressure at transducer P7. The maximum gauge pressures increase with increasing length of the vertical pipe steel along the pipeline, with gauge pressure values between 1,49 barg and 1,59 barg. In addition, higher vertical steel pipe lengths cause higher minimum gauge pressures along the pipeline, which occur at the beginning of the transient stage. At transducer P7, the highest minimum gauge pressure is $-0,03$ barg. The minimum gauge pressure occurs at the upstream end. The minimum and maximum gauge pressures occur simultaneously during the transient stage. When the air-water front reaches a particular point during the transient stage, the gauge pressure is the driving gauge pressure.

6.5   Conclusions

A mathematical model for accurately determining the emptying process in a pipeline using pressurized air was developed in this paper. It can be used for several pipeline configurations by changing only the gravity term.

The proposed model was validated using data recorded by Laanearu et al. (2012) for a PVC-steel pipeline that was 271,6 m long with a 232 mm internal diameter. The model effectively simulated measurements of the transient water flow and gauge pressure along the pipeline for 9 runs.

The results indicate that the proposed model for analyzing the emptying process can better predict both the water flow and gauge pressure along a pipeline than previous models.

The length of the vertical pipe significantly influences the results. The time for drainage decreases as the vertical steel pipe length increases, possibly because the water flow increases. However, it should be considered that the maximum and minimum gauge pressures will also increase, putting the system at risk. Considering this, it is recommended to select a suitable length for the vertical steel pipe to reduce the probability of system failure.

Notation

The following symbols are used in this paper:

$$
\begin{aligned}
A &= \text{cross-sectional area of pipe } (m^2); \\
D &= \text{internal pipe diameter (m);} \\
f &= \text{pipe wall friction coefficient } (-); \\
k(\theta) &= \text{minor-loss coefficient } (-); \\
g &= \text{gravity acceleration } (m/s^2); \\
h_s &= \text{length of the emptying column at the vertical pipe } (m); \\
h_{s,0} &= \text{height of the vertical pipe } (m); \\
Le &= \text{length of the emptying column } (m); \\
Le_0 &= \text{initial length of the emptying column } (m); \\
Le_P &= \text{length of the emptying column until point P } (m); \\
L_T &= \text{pipe length } (m); \\
p_1 &= \text{driving gauge pressure } (Pa); \\
p_{atm} &= \text{atmospheric pressure } (Pa); \\
p_P &= \text{gauge pressure at point } P \ (Pa); \\
Q &= \text{discharge } (m^3/s); \\
t &= \text{time } (s); \\
v &= \text{water velocity } (m/s); \\
x &= \text{axial coordinate } (m); \\
x_{i,0} &= \text{initial air-water front coordinate } (m); \\
\rho_w &= \text{water density } (kg/m^3);
\end{aligned}
$$

Chapter 7

Emptying operation of water supply networks

This chapter is extracted from the paper:

Emptying operation of water supply networks
Coauthors: Coronado-Hernández O. E; Fuertes-Miquel, V. S; Angulo-Hernández, F. N.
Journal: Water ISSN 2073-4441.
2017 Impact Factor: 2.069. Position JCR 34/90 (Q2). Water Resources.
State: Published. 2018. Volume 10; Issue 1; 22. DOI: 10.3390/w10010022.

7.2   Introduction

Trapped air can be introduced in water supply networks affecting the water behavior during operational stages (Apollonio et al., 2016; Balacco et al., 2015). If trapped air is compressed, so pressure surges occur (Izquierdo et al., 1999; Zhou et al., 2013a); in contrast, when it is expanded then negative pressures are reached (Coronado-Hernández et al., 2017b). To avoid these situations, air valves should be installed along pipe systems since these devices relieve transient events trough expelled/admitted air (AWWA, 2001; Ramezani et al., 2015).

Researchers have analyzed the consequences of transient events in pipelines in the following situations: during the filling process (Covas et al., 2010; Martins et al., 2015; Izquierdo et al., 1999; Zhou et al., 2013a), at pumping stations (Pozos-Estrada et al., 2016; Abreu et al., 1999), analyzing the propagation of air pockets (Escarameia, 2005), in valve closures (Abreu et al., 1999), and by using protection devices (Wylie and Streeter, 1993; Bianchi et al., 2007). The authors have developed a mathematical model to analyze the emptying process using experimental facilities (Coronado-Hernández et al., 2017b; Fuertes-Miquel et al., 2018b) which was validated for a single pipe at the Universitat Politècnica de València, Spain (Fuertes-Miquel et al., 2018b), and for a pipeline of irregular profile at the University of Lisbon, Portugal (Coronado-Hernández et al., 2017b). The main hydraulic and thermodynamic variables (absolute pressure, water velocity, and length of the emptying columns) were measured during the experiments. Figure 7.1 presents the used experimental facilities.

However, there is a lack regarding to the application of the emptying process in actual pipelines which can be used for engineers to plan the process. In water supply networks are required emptying maneuvers (Coronado-Hernández et al., 2017b). Knowing the behavior of transient flow both water and air phase is important to simulate the emptying process; this is because the water flow will be replaced by the air flow. The emptying process starts when drain valves are opened, and immediately trapped air pockets are expanded and values of subatmospheric pressure are reached. Water velocities should be varied from 0,3 to 0,6 m/s to do an appropriate operation (Ramezani et al., 2015), and air valves for vacuum protection should be installed adequately along of pipelines (Coronado-Hernández et al., 2017b; Fuertes-Miquel et al., 2018b). Air valves should open to admit an adequate quantity of air flow into the installation with a similar ratio than water flow in order to prevent subatmospheric pressure conditions. A maximum differential pressure of 34 kPa (5 psi) is recommended during the transient event (AWWA, 2001). However, if air valves are not installed or failed during the process, then the hydraulic system can collapse depending on the values of the subatmospheric pressure, and the installation conditions (e.g soil in natural conditions, type of backfill, and cover depth).

Pipe manufacturers suggest to select the stiffness pipe as a function of the lowest value of subatmospheric pressure and others conditions of the pipe installations. Engineers select the stiffness class pipe based on their experiences. Another design feature is the selection of the air valve size for vacuum protection. The bigger the air valve size, the greater protection of the system is reached (AWWA, 2001). If air valves are improperly sized, the hydraulic event can damage not only the pipeline but also protection devices (Ramezani et al., 2016;

(a) Single pipe (Universitat Politècnica de València, Spain)

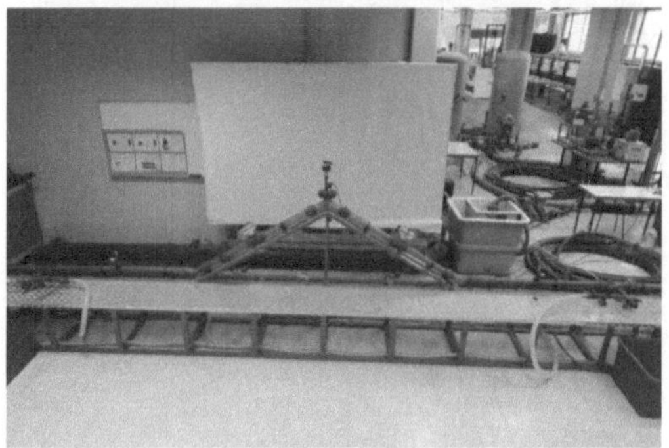

(b) Pipeline of irregular profile (University of Lisbon, Portugal)

Figure 7.1: Experimental facilities to validate the emptying process

Apollonio et al., 2016). Drain valves maneuvers should be considered during the emptying process.

This Chapter presents the mathematical model developed by the authors (Coronado-Hernández et al., 2017b; Fuertes-Miquel et al., 2018b) to analyze the emptying process, which was applied to the Ciudad del Bicentenario pipeline located in Cartagena, Bolívar Department, Colombia, in order to show the behavior of hydraulic and thermodynamic variables. The mathematical model can be used for engineers to conduct a water emptying process in pipelines in the design and planning stages to avoid the collapse of pipe systems.

7.3   Pipeline description

The Ciudad del Bicentenario pipeline is located on the southern of the city of Cartagena, Bolívar Departament, Colombia (see Figure 7.2). The company Aguas de Cartagena (Cartagena, Colombia) provided the entire information regarding to the pipeline (Aguas de Cartagena, 2013).

Figure 7.3 shows the scheme of the analyzed pipeline which is located between chainages from $K2 + 360$ m to $K2 + 940$ m. It is configured as a Glass-Reinforced Plastic (GRP) pipeline with nominal diameter of 1000 mm, 580 m long, pressure class of PN 10, and stiffness class of SN 5000 (Flowtite, 2008). Slope branches vary in a range from 0% to 6,9%. The system presents the emptying column No. 1 (chainages from $K2 + 360$ m to $K2 + 620$ m), and the emptying column No. 2 (chainages from $K2 + 940$ m to $K2 + 620$ m).

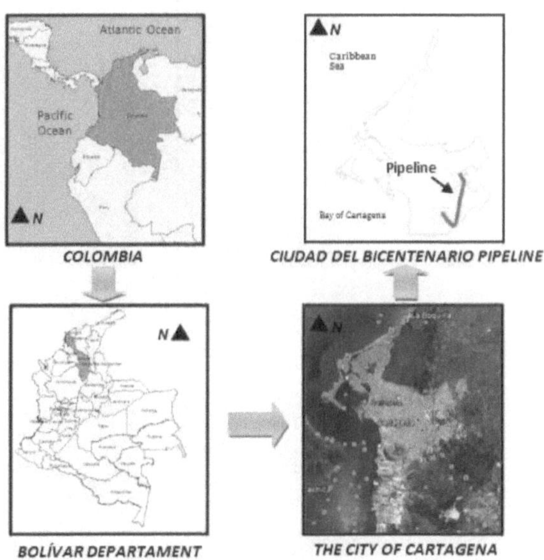

Figure 7.2: Location of the Ciudad del Bicentenario pipeline

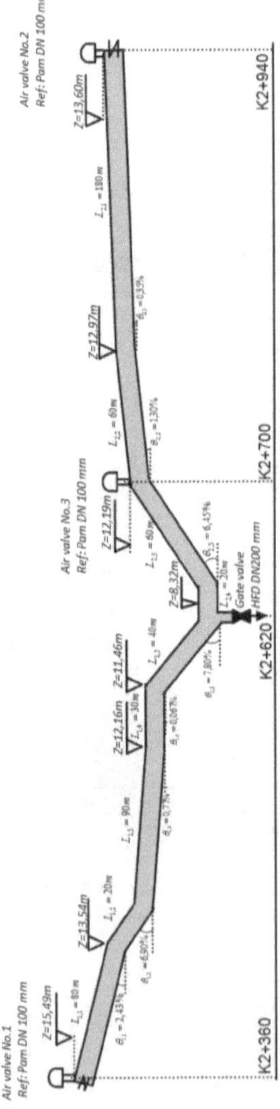

Figure 7.3: Scheme of the Ciudad del Bicentenario pipeline

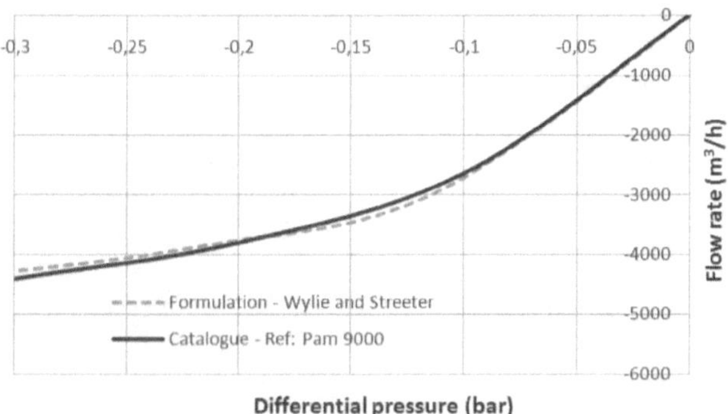

Figure 7.4: Characteristics of the air valve - Type 9000 Pam manufacturer

Three air valves were installed along of the pipeline with nominal diameter of 100 mm (Type 9000 - Pam manufacurer). They are located at chainages $K2+360$ m, $K2+700$ m, and $K2+940$ m (see Figure 7.3). Air valves should work in subsonic flow ($-0,3$ bar $< \Delta P < 0$) during the emptying process (AWWA, 2001; Ramezani et al., 2016). Using the formulation presented by Wylie and Streeter Wylie and Streeter (1993), the inflow discharge coefficient (C_{adm}) was calibrated with a value of 0,68 to get similar values to the data provided by Pam manufacturer (Saint-Gobain, 2009) as shown in Figure 7.4.

A Ductile Cast Iron (DCI) gate valve with nominal diameter of 200 mm was installed at chainage $K2+620$ m to drain the hydraulic system, with a resistance coefficient (R_v) of 0,67 ms^2/m^6. A synthetic maneuver was considered with an opening time of 200 s.

7.4   Application of the proposed model

The emptying process was modeled based on formulations developed by the authors considering as the most critical condition when the pipeline is completely filled since it produces the lowest value of the subatmospheric pressure (Coronado-Hernández et al., 2017b; Fuertes-Miquel et al., 2018b). The mathematical model uses the following formulations: i) rigid model to represent the water phase behavior (Fuertes-Miquel et al., 2016, 2018b; Coronado-Hernández et al., 2017b; Izquierdo et al., 1999); ii) a piston flow to describe the air-water interface (Zhou et al., 2013a,b; Coronado-Hernández et al., 2017b; Izquierdo et al., 1999); iii) a polytropic equation to describe the air phase behavior (Martin, 1976; Graze et al., 1996; Leon et al., 2010); (iv) the continuity equation of the air pocket (Coronado-Hernández et al., 2017b; Fuertes-Miquel et al., 2018b); and (v) the air valve characterization (Wylie and

Streeter, 1993) to quantify the admitted air volume.

7.4.1 Equations

The corresponding equations of the hydraulic system are:

1. Mass oscillation equation applied to the emptying column 1

$$\frac{dv_{w,1}}{dt} = \frac{p_1^* - p_{atm}^*}{\rho_w L_{e,1}} + g\frac{\Delta z_{e,1}}{L_{e,1}} - f\frac{v_{w,1}|v_{w,1}|}{2D} - \frac{R_v g A^2 (v_{w,1} + v_{w,2})|v_{w,1} + v_{w,2}|}{L_{e,1}} \tag{7.1}$$

2. Emptying column 1 position

$$\frac{dL_{e,1}}{dt} = -v_{w,1} \rightarrow L_{e,1} = L_{e,1,0} - \int_0^t v_{w,1}dt \tag{7.2}$$

3. Mass oscillation equation applied to the emptying column 2

$$\frac{dv_{w,2}}{dt} = \frac{p_2^* - p_{atm}^*}{\rho_w L_{e,2}} + g\frac{\Delta z_{e,2}}{L_{e,2}} - f\frac{v_{w,2}|v_{w,2}|}{2D} - \frac{R_v g A^2 (v_{w,1} + v_{w,2})|v_{w,1} + v_{w,2}|}{L_{e,2}} \tag{7.3}$$

4. Emptying column 2 position

$$\frac{dL_{e,2}}{dt} = -v_{w,2} \rightarrow L_{e,2} = L_{e,2,0} - \int_0^t v_{w,2}dt \tag{7.4}$$

5. Evolution of the air pocket 1

$$\frac{dp_1^*}{dt} = \frac{p_1^* k}{(L_1 - L_{e,1})}\left(\frac{\rho_{a,cn}Q_{a,cn,1}}{A\rho_{a,1}} - v_{w,1}\right) \tag{7.5}$$

6. Continuity equation of the air pocket 1

$$\frac{d\rho_{a,1}}{dt} = \frac{\rho_{a,nc}Q_{a,nc,1} - v_{w,1}A\rho_{a,1}}{A(L_1 - L_{e,1})} \tag{7.6}$$

7. Air valve 1 characterization

$$Q_{a,nc,1} = C_{adm}A_{adm}\sqrt{7p_{atm}^*\rho_{a,nc}\left[\left(\frac{p_1^*}{p_{atm}^*}\right)^{1.4286} - \left(\frac{p_1^*}{p_{atm}^*}\right)^{1.714}\right]} \tag{7.7}$$

8. Evolution of the air pocket 2

$$\frac{dp_2^*}{dt} = \frac{p_2^* k}{(L_2 - L_{e,2})}\left(\frac{\rho_{a,cn}Q_{a,cn,2} + \rho_{a,cn}Q_{a,cn,3}}{A\rho_{a,2}} - v_{w,2}\right) \tag{7.8}$$

9. Continuity equation of the air pocket 2

$$\frac{d\rho_{a,2}}{dt} = \frac{\rho_{a,nc}Q_{a,nc,2} + \rho_{a,nc}Q_{a,nc,3} - v_{w,2}A\rho_{a,2}}{A(L_2 - L_{e,2})} \tag{7.9}$$

10. Air valve 2 characterization

$$Q_{a,nc,2} = C_{adm}A_{adm}\sqrt{7p^*_{atm}\rho_{a,nc}\left[\left(\frac{p^*_2}{p^*_{atm}}\right)^{1.4286} - \left(\frac{p^*_2}{p^*_{atm}}\right)^{1.714}\right]} \tag{7.10}$$

11. Air valve 3 characterization

$$Q_{a,nc,3} = C_{adm}A_{adm}\sqrt{7p^*_{atm}\rho_{a,nc}\left[\left(\frac{p^*_2}{p^*_{atm}}\right)^{1.4286} - \left(\frac{p^*_2}{p^*_{atm}}\right)^{1.714}\right]} \tag{7.11}$$

It is very important to note that air valves 1 and 2 are working all time because are located at the ends of the pipeline. However, air valve 3 only is working when the air pocket passes through of the position $K2 + 700$ m.

A summary, at set of 11 algebraic-differential equations describes the hydraulic and thermodynamic behavior of the emptying process which can solved for the 11 unknowns variables: $v_{w,1}$, $v_{w,2}$, $L_{e,1}$, $L_{e,2}$, p^*_1, p^*_2, $\rho_{a,1}$, $\rho_{a,2}$, $v_{a,cn,1}$, $v_{a,cn,2}$, and $v_{a,cn,3}$.

7.4.2 Initial and boundary conditions

Considering the system is at rest ($t = 0$), then initial conditions are: $v_{w,1}(0) = 0$, $v_{w,2}(0) = 0$, $L_{e,1}(0) = 259$ m, $L_{e,2}(0) = 319$ m, $p^*_1(0) = 101325$ Pa, $p^*_2(0) = 101325$ Pa, $\rho_{a,1}(0) = 1,205$ kg/m^3, $\rho_{a,2}(0) = 1,205$ kg/m^3, $v_{a,cn,1}(0) = 0$, $v_{a,cn,2}(0) = 0$, and $v_{a,cn,3}(0) = 0$. The upstream boundary conditions are presented at chainages $K2 + 360$ m and $K2 + 940$, where the air pocket 1 and 2 are at atmospheric conditions (p^*_{atm}). The downstream boundary condition is given by the opening of the gate valve (located at chainage $K2 + 620$) where the atmospheric pressure (p^*_{atm}) is presented by the free discharge.

7.4.3 Gravity term

The gravity term of the emptying column 1 is computed by:

If $L_{1,1} + L_{1,2} + L_{1,3} + L_{1,4} + L_{1,5} \geq L_{e,1} > L_{1,2} + L_{1,3} + L_{1,4} + L_{1,5}$, then:

$$\begin{aligned}
\frac{\Delta z_1}{L_{e,1}} = {} & \frac{L_{1,2}\sin\theta_{1,2} + L_{1,3}\sin\theta_{1,3} + L_{1,4}\sin\theta_{1,4} + L_{1,5}\sin\theta_{1,5}}{L_{e,1}} \\
& + \left(1 - \frac{L_{1,2} + L_{1,3} + L_{1,4} + L_{1,5}}{L_{e,1}}\right)\sin\theta_{1,1}
\end{aligned} \tag{7.12}$$

If $L_{1,2} + L_{1,3} + L_{1,4} + L_{1,5} \geq L_{e,1} > L_{1,3} + L_{1,4} + L_{1,5}$, then:

$$\frac{\Delta z_1}{L_{e,1}} = \frac{L_{1,3} \sin \theta_{1,3} + L_{1,4} \sin \theta_{1,4} + L_{1,5} \sin \theta_{1,5}}{L_{e,1}} + \left(1 - \frac{L_{1,3} + L_{1,4} + L_{1,5}}{L_{e,1}}\right) \sin \theta_{1,2} \tag{7.13}$$

If $L_{1,3} + L_{1,4} + L_{1,5} \geq L_{e,1} > L_{1,4} + L_{1,5}$, then:

$$\frac{\Delta z_1}{L_{e,1}} = \frac{L_{1,4} \sin \theta_{1,4} + L_{1,5} \sin \theta_{1,5}}{L_{e,1}} + \left(1 - \frac{L_{1,4} + L_{1,5}}{L_{e,1}}\right) \sin \theta_{1,3} \tag{7.14}$$

If $L_{1,4} + L_{1,5} \geq L_{e,1} > L_{1,5}$, then:

$$\frac{\Delta z_1}{L_{e,1}} = \frac{L_{1,5} \sin \theta_{1,5}}{L_{e,1}} + \left(1 - \frac{L_{1,5}}{L_{e,1}}\right) \sin \theta_{1,4} \tag{7.15}$$

If $L_{1,5} \geq L_{e,1} > 0$, then:

$$\frac{\Delta z_1}{L_{e,1}} = \sin \theta_{1,5} \tag{7.16}$$

The gravity term of the emptying column 2 is computed by:
If $L_{2,1} + L_{2,2} + L_{2,3} + L_{2,4} \geq L_{e,2} > L_{2,2} + L_{2,3} + L_{2,4}$, then:

$$\frac{\Delta z_2}{L_{e,2}} = \frac{L_{2,2} \sin \theta_{2,2} + L_{2,3} \sin \theta_{2,3}}{L_{e,2}} + \left(1 - \frac{L_{2,2} + L_{2,3} + L_{2,4}}{L_{e,2}}\right) \sin \theta_{2,1} \tag{7.17}$$

If $L_{2,2} + L_{2,3} + L_{2,4} \geq L_{e,2} > L_{2,3} + L_{2,4}$, then:

$$\frac{\Delta z_2}{L_{e,2}} = \frac{L_{2,3} \sin \theta_{2,3}}{L_{e,2}} + \left(1 - \frac{L_{2,3} + L_{2,4}}{L_{e,2}}\right) \sin \theta_{2,2} \tag{7.18}$$

If $L_{2,3} + L_{2,4} \geq L_{e,2} > L_{2,4}$, then:

$$\frac{\Delta z_2}{L_{e,2}} = \left(1 - \frac{L_{2,4}}{L_{e,2}}\right) \sin \theta_{2,3} \tag{7.19}$$

If $L_{2,4} \geq L_{e,2} > 0$, then:

$$\frac{\Delta z_2}{L_{e,2}} = 0 \tag{7.20}$$

7.5   Results and discussion

The resolution of these set of differential-algebraic equations (equations 7.1-7.11) was conducted using Matlab. In order to simulate the most critical situations, small air pocket sizes

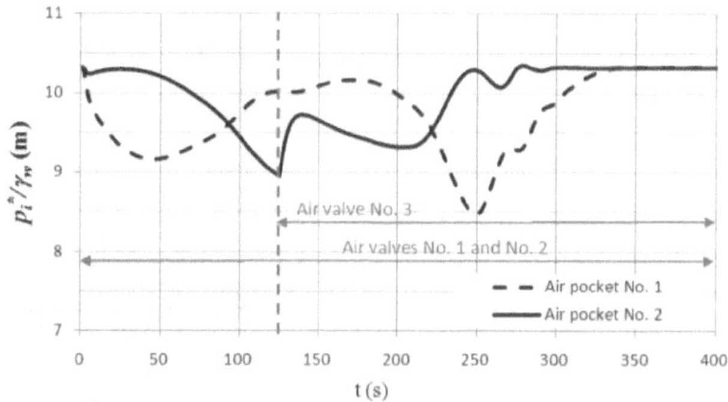

Figure 7.5: Evolution during the hydraulic event of the absolute pressure of the air pocket

were considered. However, to avoid a numerical problem a minimum air pocket length of 1 mm was established in water columns 1 and 2 (Coronado-Hernández et al., 2017b). A friction factor of 0,011 was used considering an absolute roughness k_s of 0,029 mm; and a polytropic coefficient (k) of 1,2.

7.5.1 Absolute pressure and density of the air pocket

In the air pocket 1 the minimum subatmospheric pressure head occurs at 249,4 s with a value of 8,48 m (negative pressure of $-1,85$ m) which is reached after of the total opening of the gate valve (see Figure 7.5). On the other hand, in the air pocket 2 the minimum value was 8,96 m (negative pressure of $-1,37$ m) at 125,1 s which occurs because the quantity of admitted air by air valve 2 is not enough. However, when the air valve 3 starts to admit air into the system then the absolute pressure pattern rises in the air pocket 2. At the end of the hydraulic event, emptying columns 1 and 2 reach the atmospheric conditions (p_{atm}^*). The minimum values of the subatmospheric pressure were checked by technical personnel during the emptying process of the pipeline.

The evolution of the air pocket density exhibits a parallel behavior to the absolute pressure pattern as expected for the polytropic model (equations 7.5 or 7.8) and the formulation of ideal gas law (Zhou et al., 2013b) ($p^* = \rho_a R T_a$).

7.5.2 Length of the emptying columns

Transient phenomenon shows how the lengths of the emptying columns 1 and 2 are drained during the entire hydraulic event, as a consequence of admitted air by air valves into the

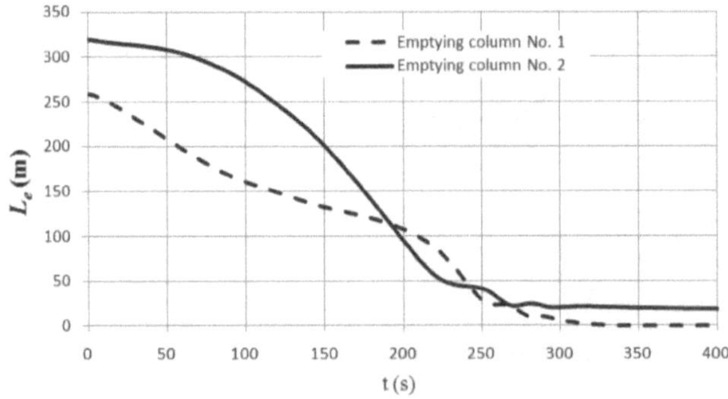

Figure 7.6: Evolution during the hydraulic event of the length of the emptying columns

system (see Figure 7.6). The emptying column 1 is drained completely at 338,9 s, whereas the emptying column 2 cannot drain completely due to horizontal branch of 20 m long located at the end, which drains more slowly and part of the water column can remain inside of the installation.

7.5.3 Water and air flow of emptying columns

Figure 7.7 presents the evolution of the water and air flow for the emptying column 1 where during the first 157,2 s the water volume drained is practically the same as the air volume admitted by the air valve 1. During this time, the subatmospheric pressure head values are higher than 9,17 m. The minimum value of 8,48 m (at 249,3 s) of the absolute pressure head is reached after the total opening of the gate valve ($T_m = 220$ s). The water flow surpasses the air flow generating the trough of the subatmospheric pressure aforementioned. At the end, some oscillations occur because the air flow is greater than water flow.

Figure 7.8 shows the behavior of the water and air flow in the emptying column 2. At the beginning of the hydraulic event (from 0 s to 54 s), the air valve 2 can introduce adequately an enough quantity of air volume, where the water and air flow are similar with values of 325, 6 l/s and 311, 4 l/s, respectively. Then, from 54 s to 125 s the water flow surpasses the air flow. The minimum value of 8,96 m of subatmospheric pressure head is reached at 125 s, when the air valve 3 starts to relief the subatamospheric pressure occurrence avoiding the risk to collapse of the installation at least until 132,7 s. At 201,2 s, air valves 2 and 3 cannot admit the required air volume generating another trough of subatmospheric pressure of 9,32 m. At the end, some oscillations occur similar to the emptying column 1.

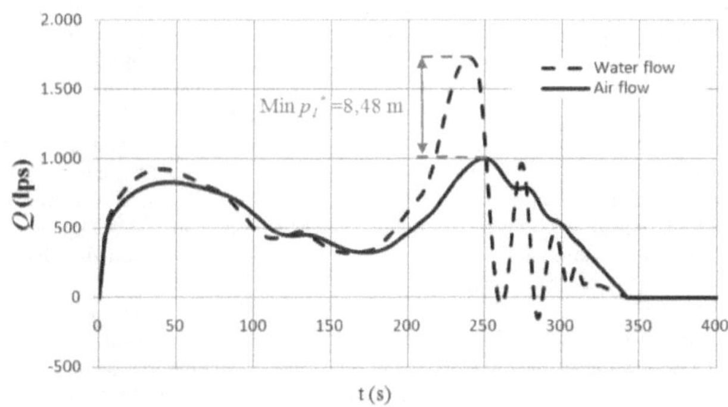

Figure 7.7: Evolution of the water and air flow in the emptying column 1

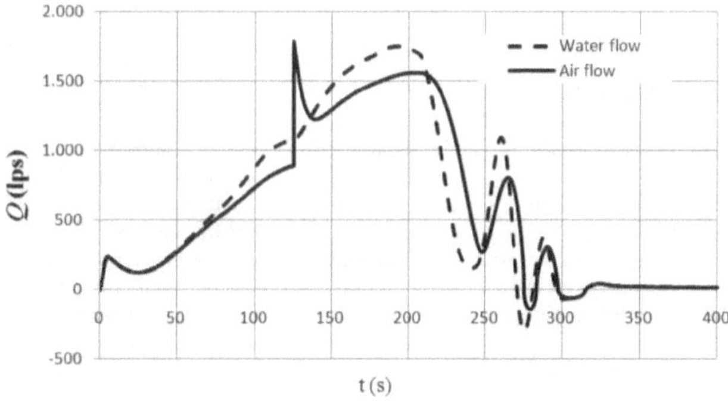

Figure 7.8: Evolution of the water and air flow in the emptying column 2

7.5.4   Risk of pipeline collapse

According to pipe manufacturer the stiffness class of a pipeline should be selected considering two parameters: (i) burial conditions, which include the soil in natural conditions, the type of backfill, and the cover depth; and (ii) the minimum value of the subatmospheric pressure. The analyzed GRP pipeline (Amiantit Pipe Systems manufacturer) of nominal 1000 mm diameter and stiffness class of SN 5000 can support a subatmospheric pressure value of 7,78 m (negative pressure of $-0{,}25$ bar or $-2{,}55$ m) considering a cover depth of 10 m, and a typical type of backfill and a soil in natural conditions. During the emptying process, the minimum value of 8,48 m (or $-0{,}18$ bar) is reached at 249,3 s with a depth of 5,75 m according to construction drawing which is lower than the values aforementioned. As a consequence, there is not risk of collapse of the installation checking that air valves and the operation of the gate valve are adequate.

Figure 7.9 presents the minimum values of subatmospheric pressure depending on the failure of air valves 1, 2, or 3. Four zones are identified. Zone I is presented when both the air valves are working or the air valve No.3 fails, then the hydraulic system is completely protected of the subatmospheric pressure occurrence and there is no risk of collapse. Zones II and IV are characterized for the failure of the air valves located at the ends. If the air valve No. 1 fails, then the air pocket 1 can cause the collapse of branch pipe 1 with a value of subatmospheric pressure head of 3,20 m; and if the air valve No.2 fails, then the minimum value of subatmospheric pressure head is 4,90 m in branch pipe 2. The most critical condition is presented in Zone III, when the air valves fail. Branch pipes 1 and 2 can collapse with values of subatmospheric pressure head of 2,15 m and 4,18 m, respectively. A well maintenance of air valves is crucial to minimize risk of collapse of pipelines during the emptying process.

7.6   Conclusions

This research presents both the mathematical model to simulate the emptying operation in water supply networks and the application to the Ciudad del Bicentenario pipeline located in Cartagena, Colombia. The analyzed pipeline has a nominal diameter of 1000 mm and 580 m long. The entire information about the pipeline was supplied by Aguas de Cartagena which operates the water supply network in Cartagena, Colombia. The mathematical model gives important information regarding the main hydraulic and thermodynamic variables (water velocities, length of the water columns, air densities, absolute pressure of air pockets, and air flows).

Regarding to the application of the mathematical model, the following conclusions can be drawn:

- The mathematical model can be used for computing air valve sizes, maneuvering drain valves, and knowing the drainage time of pipelines to prevent the collapse of the hydraulic system.

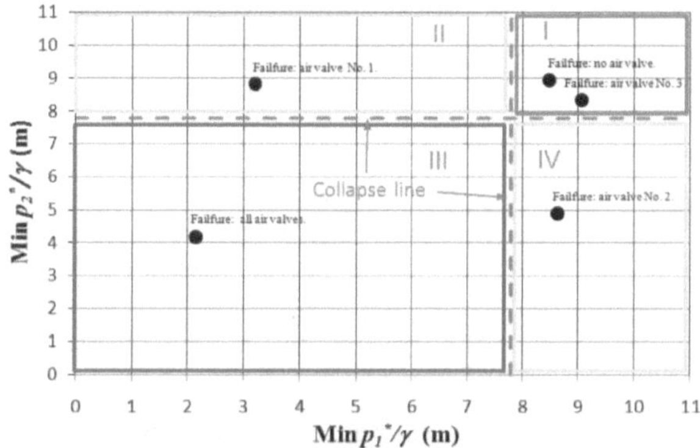

Figure 7.9: Sensitivity analysis regarding to the failure of air valves

- The minimum value of subatmospheric pressure is one of the most critical situations during the emptying process which is adequately predicted by the mathematical model.

- Air valves should be selected appropriately along of pipe systems admitting the required air to avoid subatmospheric pressure conditions. If air valves have not been well sized, then extreme negative pressures are reached which can produce the collapse of pipe systems. The mathematical model can be used to check the air valves behavior during the emptying maneuvers in actual installations.

- The Ciudad del Bicentenario pipeline can resist the minimum value of subatmospheric pressure showing there is no risk of collapse since the total volume of admitted air is similar to the volume of drained water. Both air valve sizes and the maneuver of the gate valve were adequately designed, so can be used to empty the pipeline without risk of collapse. Annual maintenance is required for the air valves located at the ends of the installation since a failure of these devices can produce the collapse of branch pipes according to the sensitivity analysis.

- Horizontal branches in pipelines are not recommended because part of the water column can remain inside of the installation and free surface flow is presented, generating a slow drainage of the system.

This research only analyzed the risk of pipeline collapse during the emptying process. However, others sources of subatmospheric pressure occurrence should be analyzed (e.g pumps' stoppages).

Notation

A	cross-sectional area of pipe (m^2);
A_{adm}	cross sectional area of the air valves (m^2);
C_{adm}	inflow discharge coefficient of the air valves (–);
D	internal pipe diameter (m);
f	friction factor (–);
g	gravity acceleration (m/s^2);
$L_{e,1}$ and $L_{e,2}$	length of the emptying columns 1 and 2, respectively (m);
L_1 and L_2	total length of the pipes 1 and 2, respectively (m);
$L_{i,j}$	branch length i,j (m);
k	polytropic coefficient (–);
p_1^* and p_2^*	absolute pressure of the air pockets 1 and 2, respectively (Pa);
p_{atm}^*	atmospheric pressure (Pa);
R	gas constant (J/kg/$^\circ$K);
R_v	resistance coefficient of the gate valve (s^2/m^5);
t	time (s);
T_a	absolute temperature of the air ($^\circ$K)
$Q_{a,nc,1}$, $Q_{a,nc,2}$, and $Q_{a,nc,3}$	admitted air flow by air valves 1, 2, and 3, respectively (m^3/s);
$v_{w,1}$ and $v_{w,2}$	water velocity of the emptying columns 1 and 2, respectively (m/s);
$\rho_{a,1}$ and $\rho_{a,2}$	air density of the air pockets 1 and 2, respectively (kg/m^3);
$\rho_{a,nc}$	air density in normal conditions (kg/m^3);
ρ_w	water density (kg/m^3);
$\Delta z_{e,1}$ and $\Delta z_{e,2}$	difference elevation of the water columns 1 and 2, respectively;
$\theta_{i,j}$	branch slope i,j

Chapter 8

Discussion

This chapter presents the doctoral book discussion. All the objectives established have been addressed to allow a better understanding of the transient phenomenon of a water emptying pipeline by considering different situations (without and with air valves).

The mathematical model can give information of the emptying processes, specifically about: (i) the main hydraulic and thermodynamic variables such as water flow, length of the water column, air absolute pressure, air density, and air flow; (ii) the risk of pipelines collapse, by checking if the stiffness class (supply by a pipe manufacturer) is appropriate to support the minimum value of subatmospheric pressure depending on the type of soil in natural conditions, the type of backfill, and the cover depth; (iii) the appropriate selection of air valves; (iv) the planning size and maneuver of drain valves; and (v) the estimation of drainage time of a pipeline.

Figure 8.1 is a flowchart that compiles information about the research stages and the proposed objectives. It also shows the interrelation among the objectives and how they were addressed in each chapter. All chapters of this document are based on papers that were written and then published as a result of the progress of this research study.

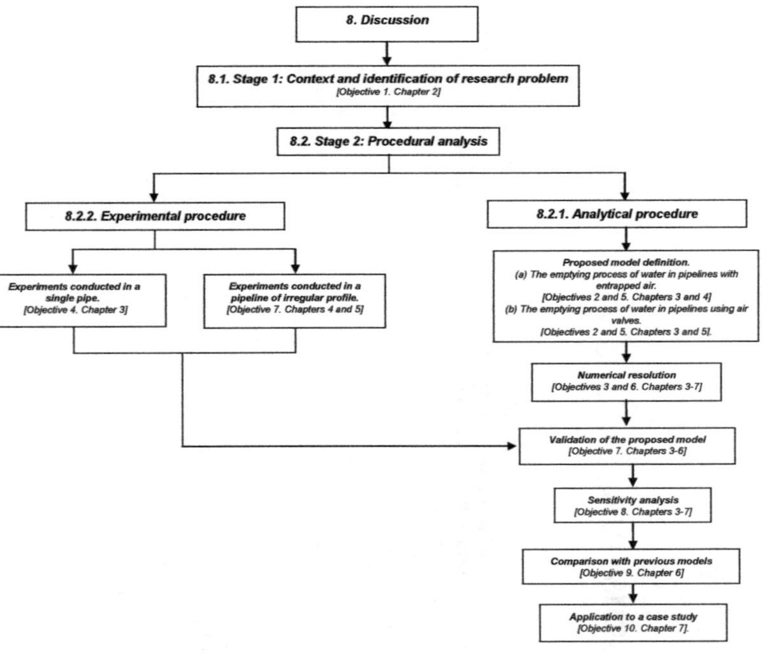

Figure 8.1: Flowchart showing research stages and proposed objectives.

The flowchart presents two research stages: (i) the Stage 1 (8.1 Context and identification of the research problem) contains a background of transient flow with trapped air in pressurized systems showing that there is a lack of knowledge related to transient phenomena during a water emptying pipeline in practical applications; and (ii) the Stage 2 (8.2 Procedural analysis) encompasses the development of a mathematical model to predict an emptying process in pressurized pipelines, which is divided in two parts (Point 8.2.1 Analytical procedure and Point 8.2.2 Experimental procedure).

8.1 Context and identification of the research problem

A literature review was conducted to detect a lack of knowledge regarding transient flow with trapped air in water supply networks, as shown in Chapter 2 (Fuertes-Miquel et al., 2018a). In this sense, none mathematical model has been developed to predict the main hydraulic and thermodynamic variables during emptying operations in water pipelines. Commonly, these operations are performed based on experiences of technical personal, where slow maneuvers are executed to avoid dangerous troughs of subatmospheric pressure. Air valves are installed along of water pipelines to prevent subamospheric pressure occurrence following typical recommendations suggested by AWWA (2001) or manufacturers. The lack of knowledge about emptying operations is identified in Chapter 2 and it is described on submitted paper:

Hydraulic modeling during filling and emptying processes in pressurized pipelines: A literature review
Coauthors: Fuertes-Miquel, V. S; Coronado-Hernández O. E; Mora-Meliá, D; Iglesias-Rey P. L..
Journal: Urban Water Journal ISSN 1573-062X.
2017 Impact Factor: 2.744. Position JCR 17/90 (Q1). Water Resources.
State: Accepted with revisions. January 2019.

To develop a reliable mathematical model to simulate emptying operations in water pipelines, the literature review covered some topics:

1. Filling and emptying procedures: filling operations have been studied by several researchers (Cabrera et al., 1992; Zhou et al., 2011b; Izquierdo et al., 1999; Fuertes-Miquel et al., 2018b) knowing the behavior under different situations; however, before this research there were not mathematical models based on physical formulations to predict emptying operations in water pipelines.

2. Mathematical models: both hydraulic (water) and thermodynamic (air) formulations were analyzed in transient flow with trapped air in order to get a robust model to predict emptying operations.

3. Methods of resolution: different methods were analyzed to solve transient flow with trapped air in order to select an adequate technique to simulate a rigid water column model to study the emptying process in pressurized pipelines.

4. Uncertainty of current models: many sources of uncertainties were identified. However, a constant friction factor and polytropic coefficient were considered to analyze emptying operations based on a literature review. On the other hand, air pocket sizes were measured during experiments in transparent pipes.

5. Design considerations: air valves selection and stiffness class need to be designed appropriately in real pipelines to avoid a pipe collapse. During experiments these conditions were checked to prevent a collapse of the two experimental facilities.

8.2   Procedural stage

This stage was divided in two parts: (i) an analytical stage where the implementation of the mathematical model to simulate a water emptying pipeline was established; and (ii) an experimental stage, which contains information about the two used experimental facilities to validate the proposed model.

8.2.1   Analytical stage

Proposed model definition

It is of great importance to explain each one of equations of the proposed model (Coronado-Hernández et al., 2018b). The mathematical model is based on physical equations showing its robustness to predict an emptying process in real pipelines. The proposed model can be used considering different conditions, such as: (i) situation No. 1 describes the condition when air valves have been installed along of pipelines, which represents the more reliability situation to protect hydraulic systems regarding to subatmospheric pressure occurrence (Fuertes-Miquel et al., 2017a); (ii) situation No. 2 describes a condition of not admitting air into pipelines, which can be produced when both air valves have failed due to maintenance problems or they have not been installed (Fuertes-Miquel et al., 2017b); (iii) situation No. 3 corresponds to the scenario for a downstream total rupture, where the greatest trough of subatmospheric pressure is attained; and (iv) situation No. 4 corresponds to an operation and a condition of drain valves (types of maneuvers, a free surface discharge or a submerged discharge).

Initially, equations to predict an emptying procedure were established for a single pipeline, which is the simplest installation (Coronado-Hernández et al., 2016). Two cases were analyzed in Chapter 3: (i) a single pipe with the upstream end closed; and (ii) a single pipe with an air valve installed in the upstream end. After that, equations were expanded to a pipeline of irregular profile considering two cases: (i) a pipeline without air valves which represents the worse case due to causing the lowest troughs of subatmospheric pressure (see Chapter 4), and (ii) a pipeline with air valves to give reliability to hydraulic installations by admitting air into systems for preventing troughs of subatmospheric pressure (see Chapter 5).

This research develops a general 1D mathematical model that can be used for analyzing the behavior of the main hydraulic and thermodynamic variables during an emptying process in a pipeline with an irregular profile and with several air valves (see Figure 8.2). An emptying process in a pipeline of irregular profile with air valves corresponds to the more complex situation to analyze by the intricacy of calculations and the configuration of real installations. Hence, an explanation of physical formulations is crucial to describe the phenomenon.

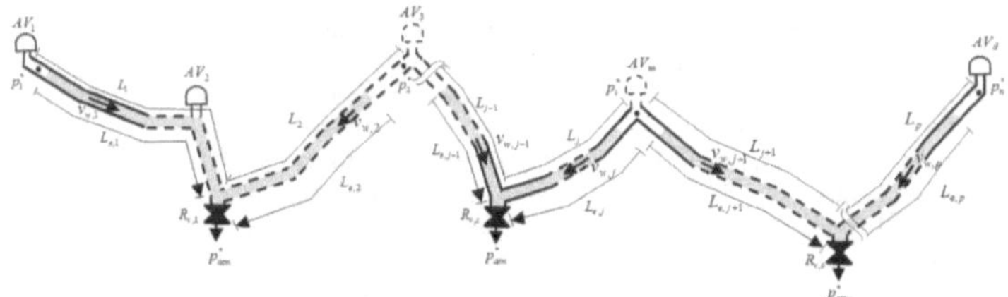

Figure 8.2: Scheme of entrapped air pockets in a pipeline with irregular profile while water empties

- Water phase relations

  Inertial models can be used to predict the behavior of a water column (Abreu et al., 1999) due to take into account system inertia. An elastic water column model (EWC) considers the elasticity of water and pipe, but is complex to implement together with air phase equations by its intricacy of calculations because a partial differential equations system is presented. This research applies a rigid water column model considering that the elasticity of air is much higher than the elasticity of pipe and water. Elastic and rigid water column models with trapped air give similar results as shown by Zhou et al. (2013a,b).

  The rigid water column model can be expressed as:

  $$H_{u,j} = H_{d,j} + h_{f,j} + h_{v,s} + \frac{L_{e,j}}{g}\frac{dv_{w,j}}{dt} \qquad (8.1)$$

  where $H_{u,j}$ and $H_{d,j}$ represent the total energy at upstream and downstream positions, respectively of the water columns j, $h_{f,j}$ is the friction losses of the water column j, $h_{v,s}$ is the local loss of the drain valve s, $L_{e,j}$ is the length of the water column j, g is the gravity acceleration, and $dv_{w,j}/dt$ is the inertial term of the water column j.

  Applying the equation 8.1 to the Figure 8.2:

$$\frac{dv_{w,j}}{dt} = \frac{p_i^* - p_{atm}^*}{\rho_w L_{e,j}} + g\frac{\Delta z_{e,j}}{L_{e,j}} - f\frac{v_{w,j}|v_{w,j}|}{2D} - \frac{R_{v,s}gA^2(v_{w,j} + v_{w,j-1})|v_{w,j} + v_{w,j-1}|}{L_{e,j}}$$

(8.2)

The gravity term $\frac{\Delta z_{e,j}}{L_{e,j}}$ produces an emptying process in pipe systems, which implies that the drainage of sloped installations are faster than horizontal systems.

- Air-water interface:

  There are several models to predict an air-water interface where is necessary to select the appropriate one depending on the nature of transient flow, the movement between liquid and gas phases, and the inertial and gravitational effects of analyzed flow (Fuertes, 2001). A piston flow model was selected because the air-water interface tends to be perpendicular to the pipe direction in pressurized systems. This assumption was checked in Chapter 3 using a single pipe of internal diameter of 42 mm with pipe slopes of 26,18° and 29,51°, in Chapter 4 and 5 using a pipeline of internal diameter of 51,4 mm of irregular profile with a slope of 30°, and in Chapter 6 was verified in a horizontal pipeline with an internal diameter of 232 mm. The equation is presented as follows:

$$\frac{dL_{e,j}}{dt} = -v_{w,j}\mathrm{d}t$$

(8.3)

  Two-phases models such as bubble flow, stratified smooth flow, stratified wave flow, and slug flow (Bousso et al., 2013) are used to simulate pipes in urban drainage systems (Vasconcelos and Wright, 2008; Vasconcelos and Marwell, 2011). The implementation of two-phases models is more complex compared to a piston flow model.

- Expansion equation for the air pocket i:

  The thermodynamic behavior of air pockets (Martins et al., 2015) is established from the first law of thermodynamics, which is known as a polytropic model (Martin, 1976; Leon et al., 2010).

$$\frac{dp_i^*}{dt} = -k\frac{p_i^*}{V_{a,i}}\frac{dV_{a,i}}{dt} + \frac{p_i^*}{V_{a,i}}\frac{k}{\rho_{a,i}}\frac{dm_{a,i}}{dt}$$

(8.4)

Based on this equation can be deduced: (i) the expansion of air pockets during the emptying process ($dV_{a,i}/dt$) produces a subatmospheric pressure occurrence; and (ii) the admitted air by air valves ($dm_{a,i}/dt$) relieves values of subatmospheric conditions (see Figure 8.3a).

If the thermodynamic process is considered isothermal ($k = 1,0$), then an expansion of air pockets produces greater values of subatmospheric pressure compared to an adiabatic process ($k = 1,4$), as shown in Figure 8.3b.

When air valves have not been installed or they are failed due to operational problems, then $dm_{a,i}/dt = 0$, and the polytropic model can be expressed as follows:

$$p_i^* V_{a,i}^k = p_{i,0}^* V_{a,i,0}^k = const. \tag{8.5}$$

Equation 8.5 shows how an expansion of air pockets can generate extreme values of subatmospheric pressure depending on the initial value of air pocket sizes, because the relation $p_i^* V_{a,i}^k$ remains constant over time.

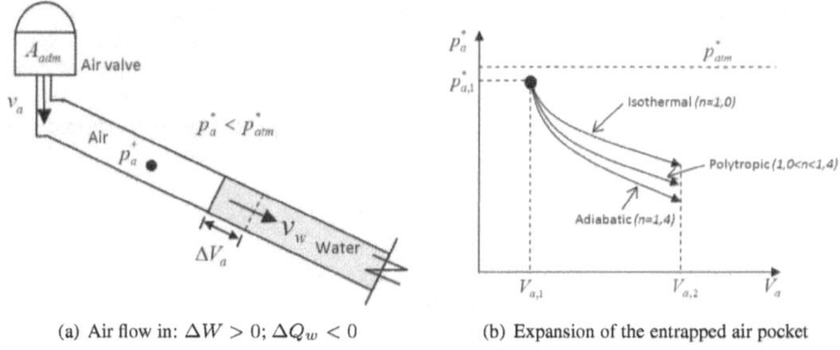

<table>
<tr><td>(a) Air flow in: $\Delta W > 0$; $\Delta Q_w < 0$</td><td>(b) Expansion of the entrapped air pocket</td></tr>
</table>

Figure 8.3: Effects of air valves behavior during the emptying process

- Air valve characterization:

  Admitted air by air valves can be expressed using different formulations. The formulation proposed by Wylie and Streeter (Wylie and Streeter, 1993; Iglesias-Rey et al., 2014) was used to represent the air admission into the system for air valves since it is based on a physical equation. Ideally an air valve m should be working in subsonic flow ($p_{atm}^* > p_i^* > 0,528p_{atm}^*$), then:

  $$Q_{a,nc,m} = C_{adm,m} A_{adm,m} \sqrt{7p_{atm}^* \rho_{a,nc} \left[\left(\frac{p_i^*}{p_{atm}^*} \right)^{1,4286} - \left(\frac{p_i^*}{p_{atm}^*} \right)^{1,714} \right]} \tag{8.6}$$

- Continuity equation of air pocket:

  This equation links the formulation of an air valve with the expansion of an air pocket, explaining how the behavior of air density inside pipelines is. This formulation is an important contribution to the literature since emptying processes with air valves can be analyzed.

This equation was developed from the continuity equation of an air pocket.

$$\frac{d\rho_{a,i}}{dt} = \frac{\rho_{a,nc}v_{a,nc,m}A_{adm,m} - (v_{w,j+1} + v_{w,j})A\rho_{a,i}}{A(L_j - L_{e,j} + L_{j+1} - L_{e,j+1})} \tag{8.7}$$

Numerical resolution

The resolution was conducted using the Simulink Library of Matlab. Simulations showed a great fit compared to experimental data, which confirms the adequacy of the used solver in Matlab. The resolution was performed using the method ODE23s, which is supported on a modified Rosenbrock formula of order 2. The resolution method was used from Chapters 3 to 7. Figure 8.4 shows the executed code to simulate an emptying process in pipe systems considering the case of a single pipe with the upstream end closed (see Chapter 3). Formulations of the rigid water column model, the polytropic model and the the air-water interface equation are presented.

Validation of the proposed model

The validation of the proposed model was performed in two experimental facilities. Based on the results, formulations to predict a water emptying pipeline were validated.

Initially, the validation was conducted in a single pipeline located at the hydraulic laboratory at the Universitat Politècnica de València (Valencia, Spain). Figure 8.5 presents how the proposed model predicts adequately the experimental data for two cases (see Chapter 3). Case No. 1 corresponds to a single pipe with the upstream end closed, where the following formulations were tested: a rigid water column model (equation 8.2), a piston flow model to represent an air-water interface (equation 8.3), and a polytropic model of an air pocket (equation 8.5); and Case No. 2 represents a single pipe with an air valve installed on the upstream end, where the air phase is described by the polytropic model with admitted air (equation 8.4), the air valve characterization (equation 8.6), and the continuity equation of an air pocket (equation 8.7).

Results are described on published paper:

Transient phenomena during the emptying process of a single pipe with water-air interaction

Coauthors: Fuertes-Miquel, V. S; Coronado-Hernández O. E; Iglesias-Rey P. L.; Mora-Meliá, D.

Journal: Journal of Hydraulic Research ISSN 0022-1686

2017 Impact Factor: 2.076. Position JCR 41/128 (Q2). Civil Engineering

State: Published. Latest Articles. 2018; DOI: 10.1080/00221686.2018.1492465

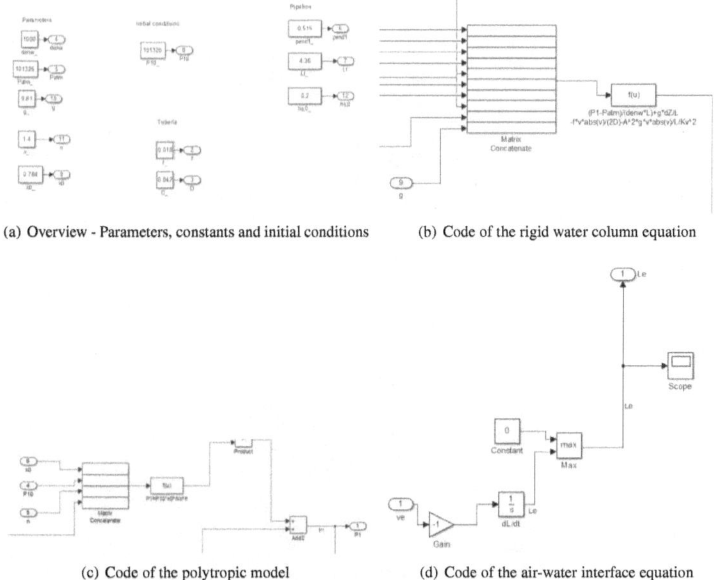

(a) Overview - Parameters, constants and initial conditions   (b) Code of the rigid water column equation

(c) Code of the polytropic model   (d) Code of the air-water interface equation

Figure 8.4: Executed code in Simulink Library of Matlab

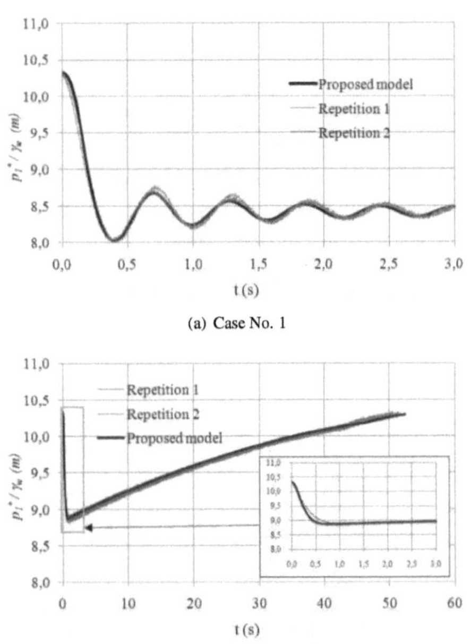

(a) Case No. 1

(b) Case No. 2

Figure 8.5: Comparison between proposed model and experiments of the absolute pressure pattern in a single pipe

Finally, the proposed model was validated in a pipeline of irregular profile (see Chapters 4 and 5). Based on the results, the proposed model can be used to predict a water emptying operation without admitted air and with air valves in real pipelines. Three situations are identified as follows: (i) Situation No. 1 represents a condition of an emptying water operation without admitted air, where an air pocket volume is distributed equally inside water columns with partial or complete opening of drain valves; (ii) Situation No. 2 is similar to the aforementioned, but an air pocket volume is not distributed equally; and (iii) Situation No. 3 is a condition of an emptying water operation using air valves. Figure 8.6 shows a comparison for the three identified situations analyzing the air absolute pressure. Situations No. 1 and No. 3 are simulated adequately by the proposed model, while Situation No. 2 cannot be totally reproduced by the mathematical model since the backflow air phenomenon is presented, which is characterized by an air entrance through drain valves. However, the proposed model is capable to predict the minimum value of absolute pressure head in all situations, which is of utmost important to check a pipeline collapse. The remaining variables (water velocity and length of water columns) presented a similar behavior regarding experimental data.

Two published papers contain the information regarding the proposed model and its validation:

Subatmospheric pressure in a water draining pipeline with an air pocket
Coauthors: Coronado-Hernández O. E; Fuertes-Miquel, V. S; Besharat, M.; and Ramos, H. M.
Journal: Urban Water Journal ISSN 1573-062X.
2017 Impact Factor: 2.744. Position JCR 17/90 (Q1). Water Resources.
State: Published. 2018. Volume 15; Issue 4. DOI: 10.1080/1573062X.2018.1475578

Experimental and Numerical Analysis of a Water Emptying Pipeline Using Different Air Valves
Coauthors: Coronado-Hernández O. E; Fuertes-Miquel, V. S; Besharat, M.; and Ramos, H. M.
Journal: Water ISSN 2073-4441.
2017 Impact Factor: 2.069. Position JCR 34/90 (Q2). Water Resources.
State: Published. 2017. Volume 9; Issue 2; 98. DOI: 10.3390/w9020098.

Sensitivity analysis

A sensitivity analysis was conducted through Chapters 3 to 7. An emptying operation can cause serious problems when an air vacuum valve has not been selected adequately because

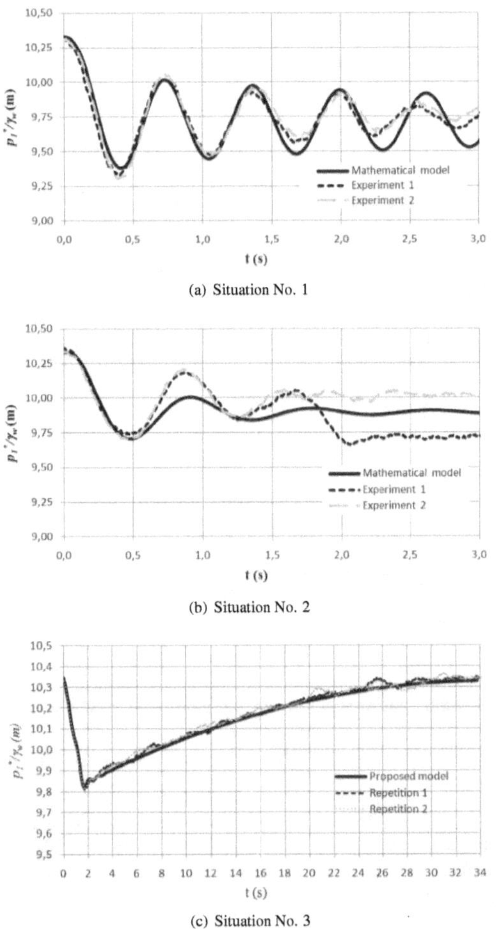

(a) Situation No. 1

(b) Situation No. 2

(c) Situation No. 3

Figure 8.6: Comparison between proposed model and experiments of the absolute pressure pattern in a pipeline of irregular profile

troughs of subatmospheric pressure are reached. The most significant parameters during the operation are air valve size, air pocket size, pipe slope, and internal pipe diameter. Friction factor and polytropic coefficient produce a despicable effect during the operation (Coronado-Hernández et al., 2017c);. Each hydraulic system should be analyzed by engineers since an expansion of air pockets can produce problems to the pipe system and its devices. Hence, the application of the mathematical model is crucial. It is recommended to perform emptying operations with slow maneuvers in valves in order to guarantee subsonic conditions in air valves to prevent extreme values of the absolute pressure. Water velocity also should be controlled according to technical manuals to avoid a system failure.

Comparison with previous models

This section presents a comparison between the proposed model and previous models. Two previous models were considered: a CFD 2D model, which is used in research to know the water behavior in complex situation; and a semi-empirical model (Tijsseling et al., 2016), which was used as a first approximation to predict a water emptying operation.

Proposed model vs. CFD model

The proposed model is a physically based $1D$ model because it uses thermodynamic equations of air pockets and hydraulic formulations of water phase; while, a CFD model simulates the transient phenomenon using the law of mass conservation, Newton's second law and law of energy conservation. In this sense, both models represent adequately the behavior of the emptying process in pressurized pipelines. To compare results the experimental facility of a single pipeline (see Chapter 3) was used analyzing the scenario without admitted air.

The proposed model can simulate rapidly the emptying procedure compared to a CFD model. As a consequence, the proposed model is more appropriate to use in practical applications. A CFD model permits to simulate the backflow air phenomenon, but this situation rarely occurs in real hydraulic systems. Figure 8.7 presents a comparison between the proposed model and a 2D CFD model in the two used experimental facilities, where initial and boundary conditions were identical. The comparison shows how both models are adequate to simulate an emptying procedure in pressurized pipelines. Appendix A shows all considerations using a 2D CFD model.

Proposed model vs. semi-empirical model

Tijsseling et al. (2016) developed a semi-empirical model to simulate an emptying process in a pipeline using pressurized air. Experiments were conducted by Laanearu et al. (2012). In this section, a comparison is presented between the proposed model and the Tijsseling's model. Chapter 6 presents details of this hydraulic installation. Figure 8.8 presents a comparison of the water flow oscillation for runs 2 and 6 (see Chapter 6) between the results obtained by the proposed model and Tijsseling's model. In all cases, the proposed model can reproduce the water flow oscillations better than Tijsseling's model. Figure 8.8b shows that

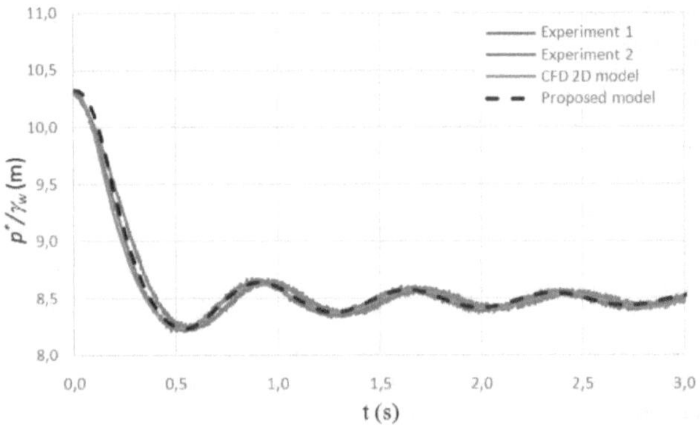

Figure 8.7: Comparison between proposed model and CFD model of the absolute pressure pattern for Test No.2 of the single pipeline (see Chapter 3)

the model developed by Tijsseling et al. (2016) cannot adequately reproduce run 6 because the water flow oscillation is too low. Figure 8.9 shows a comparison of the gauge pressure along the pipeline for run 4 at locations P1 ($x = 1{,}55$ m) and P9 ($x = 252{,}76$ m). Again, the proposed model can better predict the gauge pressure oscillations than the model developed by Tijsseling et al. (2016).

Results are presented on published paper:

Rigid Water Column Model for Simulating the Emptying Process in a Pipeline Using Pressurized Air
Coauthors: Coronado-Hernández O. E; Fuertes-Miquel, V. S; Iglesias-Rey, P.L.;
Martínez-Solano, F.J.
Journal: Journal of Hydraulic Engineering ISSN 0733-9429.
2017 Impact Factor: 2.080. Position JCR 39/128 (Q2). Civil Engineering.
State: State: Published. 2018. Volume 144; Issue 4. DOI:
10.1061/(ASCE)HY.1943-7900.0001446.

Application to a case study

One of the most important contribution is to apply the proposed model to a case study. In this sense, the Ciudad del Bicentenario pipeline located in Cartagena de Indias, Colombia, was

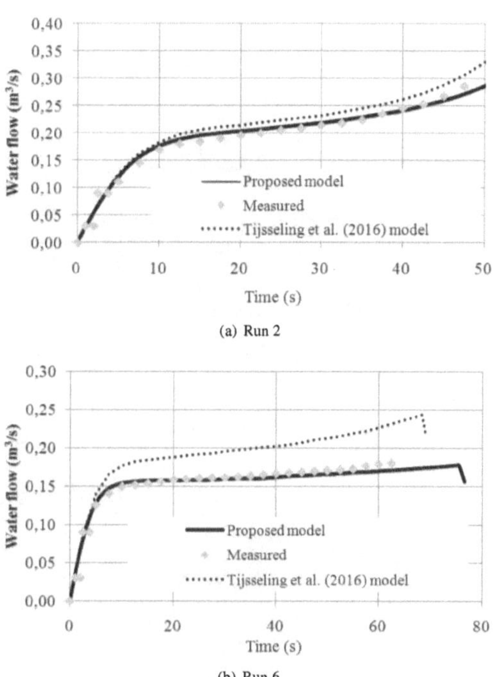

(a) Run 2

(b) Run 6

Figure 8.8: Comparison of the water flow oscillation pattern

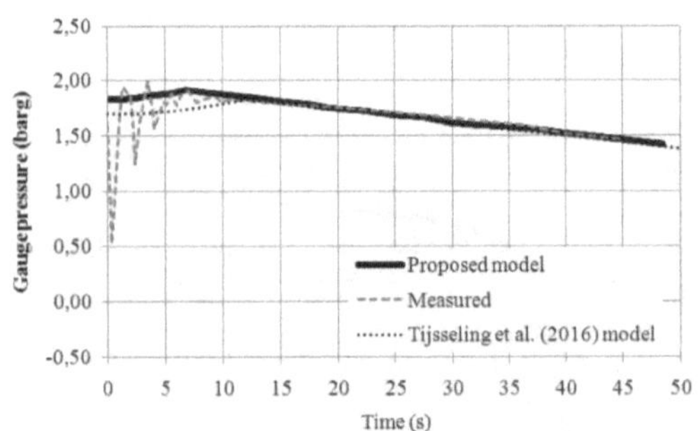

(a) Transducer 1, located at $x = 1,55$ m

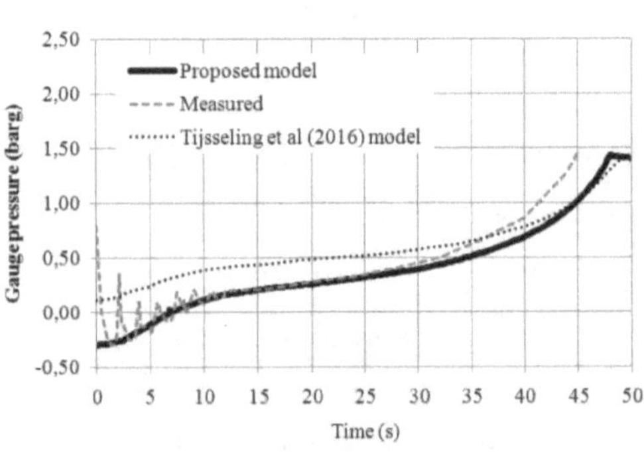

(b) Transducer 9, located at $x = 252,76$ m

Figure 8.9: Comparison of the gauge pressures along the pipeline for run 4

selected as a case study (see Chapter 7) (Coronado-Hernández et al., 2017a). The application of the mathematical model is presented for showing how is possible to simulate an emptying process in a real pipeline. Details are presented in Chapter 7, which can be used for engineers to plan an emptying operation.

Results are presented on published paper:

Emptying Operation of Water Supply Networks
Coauthors: Coronado-Hernández O. E; Fuertes-Miquel, V. S; Angulo-Hernández, F. N.
Journal: Water ISSN 2073-4441.
2017 Impact Factor: 2.069. Position JCR 34/90 (Q2). Water Resources.
State: State: Published. 2018. Volume 10; Issue 1; 22. DOI: 10.3390/w10010022.

8.2.2 Experimental stage

This stage of the research focuses on the experimental stage, which is used to validate the mathematical model to predict emptying processes in pressurized pipelines. Experimental tests were performed at the hydraulic labs of the Universitat Politècnica de València (Valencia, Spain) and the Instituto Superior Técnico, University of Lisbon (Lisbon, Portugal). At the beginning, a single pipeline was used to establish initial formulations to predict the process. After that, formulations were expanded to a pipeline of irregular profile. Figure 8.10 presents the two experimental facilities.

Experiments conducted in a single pipeline

Experiments of an emptying process applied to a single pipe (Chapter 3) were performed at the hydraulic lab of the Universitat Politècnica de València (Valencia, Spain). The installation consists in a water supply tower, a 4,16-m-long methacrylate pipe, a 0,2-m-long PVC pipe, PVC joints, two ball valves with internal diameters of 42 mm at the ends, and a free-surface basement reservoir to collect drainage water. A transducer was installed upstream to measure the air pocket pressure which was calibrated to get data using the equation: $p^*(bar) = [p^*$ (V)-4]/4. Where V represents the measured volt.

A total of 36 tests were performed for two cases: (i) Case No. 1 which represents an emptying process without admitted air, and (ii) Case No. 2 where an emptying process is performed by admitting air with orifices of 1,5 and 3,0 mm.

This section pursues two objectives:

- (i) The definition of a mathematical formulation to predict an emptying process in a single pipeline.

- (ii) the validation of the proposed model through a comparison between the computed and measured of absolute pressure.

(a) Experiments conducted in a single pipeline - Universitat Politècnica de València (Valencia, Spain)

(b) Experiments conducted in a pipeline of irregular profile - Instituto Superior Técnico, University of Lisbon (Lisbon, Portugal)

Figure 8.10: Experimental facilies used to validate the mathematical model

Experiments conducted in a pipeline of irregular profile

A second experimental facility was used to measure hydraulic and thermodynamic variables considering the main objective of the research regarding to go deeply about transient flow of an emptying process in a pipeline of irregular profile (Chapter 4 and 5). The used installation is located at the hydraulic lab at the Instituto Superior Técnico, University of Lisbon (Lisbon, Portugal).

The used installation is composed by two PVC pipes with a total length of 7.3 m and a nominal diameter of 63 mm. Two drain valves DV_1 and DV_2 are located at ends. Air valves $S050$ and $D040$ of A.R.I. manufacturer were used to simulate the case No. 2 with admitted air.

The used instrumentation is described as follows: a WIKA transducer to measure the air pocket pressure located at the high point, an Ultrasonic Doppler Velocimetry (UDV) located at the horizontal PVC pipe with a transducer of 4 MHz frequency to measure the water velocity, and a Sony Camera DSC-HX200V to measure the length of the water columns.

This section pursues three objectives:

- The complete definition of the proposed model using hydraulic and thermodynamic formulations.

- The analysis of more complex system which consists in a pipeline of irregular profile with several pipe branches, various valves, and various air valves.

- Providing additional measurements not only air absolute pressure but also water velocity and length of water column during an emptying process.

A total of 16 tests were performed for two cases: (i) Case No. 1 which corresponds to an emptying process in a water pipeline of irregular profile without admitted air, and (ii) Case No. 2 where an emptying pipeline process is performed by admitting air using air valves $S050$ and $D040$ (A.R.I. manufacturer).

Notation

A	Cross sectional area of the pipe (m^2)
$A_{adm,m}$	Cross section area of the air valve m (m^2)
$C_{adm,m}$	Inflow discharge coefficient of the air valve m (–)
D	Internal pipe diameter (m)
f	Friction factor (–)
g	Gravity acceleration (m/s^2)
$H_{d,j}$	Total energy at downstream of the water column j (m)
$H_{u,j}$	Total energy at upstream of the water column j (m)
$h_{f,j}$	Friction loss of the water column j (m)
$h_{v,s}$	Local loss of the drain valve s (m)
k	Polytropic coefficient (–)
$L_{e,j}$	Length of the emptying column j (m)
L_j	Total length of the pipe j (m)
$m_{a,i}$	Air mass of the air pocket i (kg)
p_i^*	Absolute pressure of the air pocket i (Pa)
p_{atm}^*	Atmospheric pressure (Pa)
t	Time (s)
$Q_{a,nc,m}$	Air discharge in normal conditions admitted by the air valve m (m^3/s)
$R_{v,s}$	Resistance coefficient of the drain valve s (s^2/m^5)
$V_{a,i}$	Air volume of the air pocket i (m^3)
$v_{w,j}$	Water velocity of the emptying column j (m/s)
$\Delta z_{e,j}$	Elevation difference (m)
$\rho_{a,i}$	Air density of the air pocket i (kg/m^3)
$\rho_{a,nc}$	Air density in normal conditions (kg/m^3)
ρ_a	Water density (kg/m^3)

Superscripts

*	Absolute values

Subscripts

a	Refers to air
i	Refers to air pocket
j	Refers to pipe
m	Refers to air valve
nc	Normal conditions
w	Refers to water
0	Initial condition

Chapter 9

Conclusions and recommendations

9.1　Main contributions

Entrapped air can be injected into water pipelines due to several situations, which can produce many problems because the elasticity of water and pipe is much lower than the elasticity of the air itself. This research focuses on the behavior of emptying maneuvers in water pipelines where air pockets are expanded during these operations producing values of subatmospheric pressure. Water pipelines should be emptied for maintenance, cleaning or repairs. These operations are repeated periodically, and should be considered in the design stage to avoid future problems associated to pipeline collapse.

Commonly, water pipelines present an irregular profile (and elevations) depending on the topography of the pipelines layout. Typically, air valves are chiefly located in high points and drain valves are usually located in low points. The emptying process starts when drain valves are opened. This maneuver makes air valves start to admit air into pipelines, and the drainage of water columns starts until pipes are completely emptied. However, if air valves have not been installed or they have failed due to operational and maintenance problems, then values of subatmospheric pressure can be reached. Air valves play a key role in making the pressure values not to drop drastically, which reduces the risk of having a pipeline collapse. Both situations (with and without air valves) have to be analyzed in order to optimize the maneuvers in pipelines during emptying operations. Understanding the transient flow in the emptying operation is crucial to evaluate subatmospheric pressure values, which, as previously mentioned, can produce the collapse of hydraulic systems depending on installation conditions.

First of all, a literature review was conducted regarding the transient flow with trapped air to detect a lack of knowledge about transient phenomena in water supply networks.

Over recent decades, some of the topics involved in the transient phenomenon with two phases flow have been investigated. The cavitation phenomenon has been studied both knowing the generation of vapor cavities into water (two phases flow) and how bubbles can change transient flow events. Also, the filling process in water supply networks has been analyzed by considering different scenarios and several devices. Despite these studies, there is a lack of detailed studies related to emptying processes in water pipelines. In this sense, this research investigates as deeply as possible about the analysis of transient phenomenon of a water emptying pipeline of irregular profile with several air pockets and various air valves.

The main contribution of this work is the development of a general one-dimensional (1D) mathematical model that can be used for analyzing the behavior of the main hydraulic and thermodynamic variables during emptying processes in pipelines of irregular profile. Further, the mathematical model can be used to both understand the performance of the process and compute all variables involved . The mathematical model was validated in two experimental facilities: (i) using a single pipe at the Fluids Laboratory at the Universitat Politècnica de València (Valencia, Spain), and (ii) with a pipeline of irregular profile at the University of Lisbon (Lisbon, Portugal). Comparisons between computed and measured values of the main variables show that the mathematical model can predict accurately not only extreme values, but also their patterns. The mathematical model can be used by designers and engineers to analyze emptying processes in real pipelines for different scenarios. Air absolute pressure is the main variable during emptying processes since the minimum value of subatmospheric pressure should be used to check the risk of pipelines collapse depending on the stiffness class recommended by manufacturer's.

9.2   Conclusions

The conclusions presented in this section are a compilation of the main conclusions presented at the end of each of the chapter of this research.

Chapter 2: Hydraulic modeling during filling and emptying processes in pressurized pipelines: A literature review

This chapter shows the main researches conducted about transient phenomena with entrapped air pockets. The main conclusions are:

- Emptying processes in water pipelines have been studied by few authors. Probably due to the intricacy of calculations and configurations of hydraulic systems, which allows to make new contributions in this topic.

- The rigid water column model can be used adequately for simulating a transient phenomenon with entrapped air pockets considering that the elasticity of the air is much higher than the elasticity of both the water and the pipe. The greater the air pocket size, the better results are obtained with the model. A piston flow model can be used to simulate an air-water interface.

- Air pockets were modeled using thermodynamic formulations. In spite of its simplicity, the polytropic model represented adequately the behavior of entrapped air pockets. A constant polytropic coefficient was used to simulate hydraulic events.

Chapter 3: Transient phenomena during the emptying process of a single pipe with water-air interaction

This chapter presented the mathematical model to simulate the behavior of the emptying process of a single pipe with an entrapped air pocket for two cases: (i) Case No.1 corresponded to the situation where there is no admitted air into the system; and (ii) Case No.2 corresponded to the situation where an orifice size introduces air to relive drops of absolute pressure pattern. According to the results, the expansion of an entrapped air pocket produces values of subatmospheric pressure. The main conclusions are:

- The analysis of a transient flow during emptying processes of single pipelines can be modeled using the proposed model. In Case No. 1 (a simple pipe with the upstream end closed), the proposed model was reduced to a set of three equations, which provides the evolution of the three unknowns variables (v_w, L_e, and p_1^*). In Case No. 2 (a simple pipe with an air valve installed in the upstream end), the mathematical model was reduced to a set of five equations, which provides the evolution of the five unknowns variables (v_w, L_e, p_1^*, $v_{a,nc}$, and ρ_a).

- The mathematical model was validated in an experimental facility located at the Universitat Politècnica de València (Valencia, Spain). A comparison between computed and measured of absolute pressure was conducted showing how the proposed model predicted accurately the transient flow. The validation of transient flow equations to predict emptying processes in single pipelines is crucial to continue with the development of equations in water emptying pipelines of irregular profiles.

- Slow maneuvers in drain valves are recommended as much as possible so as to admit the air required into the system.

Chapter 4: Subatmospheric pressure in a water draining pipeline with an air pocket

In this chapter a general 1D mathematical model was developed so that it can be used for modeling the emptying process of a pipeline of irregular profile without air valves. Measurements were conducted at the University of Lisbon (Lisbon, Portugal). A comparison between the measurements and the mathematical model were made. The main conclusions are:

- The emptying process in water pipelines of irregular profile with k pipes and n air pockets without air admission generated a transient flow, which can be modeled through a mathematical model with a set of $2k + n$ algebraic-differential equations. The resolution of the system equations brings the evolution of the hydraulic and thermodynamic variables such as absolute pressure, water velocity, and length of the water column.

- The mathematical model was validated in an experimental facility located at the University of Lisbon (Lisbon, Portugal), where different situations were analyzed: (i) a partial opening of drain valves with an air pocket volume distributed uniformly, (ii) a total opening of the drain valves with an air pocket volume distributed uniformly, and (iii) a partial opening of the drain valves with an air pocket volume distributed non-uniform. The mathematical model predicted accurately the behavior both pattern and extreme values of air pocket pressure, water velocity, and length of the water column for the majority of the simulations.

- The backflow air phenomenon can occur during emptying processes without admitted air depending on an opening percentage of drain valves. The mathematical model was capable of predicting the lowest value of subatmospheric absolute pressure. The phenomenon was able to minimize the subatmospheric pressure pattern by introducing air from down-stream to upstream.

- The mathematical model can be used to detect the risk of pipelines collapse in real piping systems where air valves have not been installed or when they failed due to maintenance and operational problems since the lowest value of the subatmospheric pressure was adequately represented.

Chapter 5: Experimental and numerical analysis of a water emptying pipeline using different air valves

The analysis of transient flow during emptying processes of pipelines of irregular profile with n air pockets and d air valves was developed. The main conclusions are:

- The analysis of the transient flow of an irregular profile with n entrapped air pockets, p pipes, and d air valves can be simulated with the mathematical model which presents a set of $2p + 2n + d$ algebraic-differential equations. The resolution gave information regarding the main hydraulic variables (absolute pressure of air pocket, water velocities, air density of the air pockets, lengths o the emptying columns and air flow rates).

- The mathematical model with air valves was validated in an experimental facility located at the University of Lisbon (Lisbon, Portugal). A comparison between computed and measured of air pocket pressure, water velocity and length of water columns confirmed that the mathematical model can be used to simulate the hydraulic event.

- Air valves are very important during emptying processes because they were able to reduce the values of subatmospheric pressure by introducing air into hydraulic installations. The larger the orifice size of the air valve, the lowest values of subatmospheric pressures were obtained.

Chapter 6: Rigid water column model for simulating the emptying process in a pipeline using pressurized air

This chapter was developed based on published experimental results obtained in a emptying process pipeline using pressurized air. The main conclusions are:

- The results obtained when comparing measured and computed values of water flow oscillations and gauge pressures confirmed that rigid water column model and piston flow formulations can be used to simulate emptying operations.

- The mathematical model also represented adequately the behavior of the main hydraulic variables considering different minor-loss coefficients.

- The mathematical model better predicted the behavior of emptying operations when compared to semi-empirical models.

Chapter 7: Emptying operation of water supply networks

This chapter presented the application of the mathematical model to the case study occurred in the pipeline network of the neighborhood called Ciudad del Bicentenario located in Cartagena de Indias (Colombia). All data was provided by Aguas de Cartagena (Acuacar). The main conclusions are:

- The mathematical model can be used to represent emptying processes in real pipelines as presented in the referenced chapter.

- A pipeline collapse can be checked by comparing the minimum value of subatmospheric pressure with data supplied by any pipe manufacturer depending on installation conditions.

- Air valves behavior and drain valves maneuvers can be checked during emptying operations in full-scale pipeline networks.

9.3 Future developments

During the literature review process performed during this study, a lack of detailed studies on emptying processes in pipelines and engineering application was identified.

Future studies on emptying processes can address specific topics, such as:

- Improvements of the mathematical model: the behavior of a water column separation cannot be detected by the mathematical model, then some approximations can be conducted in this topic.

- Analysis of entrapped air pockets: according to the results obtained in this study, the polytropic model reproduces adequately the behavior of entrapped air pockets into pipeline installations. This formulation is widely used considering its definition and simplicity. In some situations the polytropic model cannot predict the behavior of an air pocket; therefore, a complex thermodynamic formulation can be applied.

- Evolution of the polytropic coefficient and the friction factor: the majority of the models developed in this field considers both the polytropic coefficient and the friction factor a constant when analyzing hydraulic events. Unsteady coefficients can improve simulations during emptying processes.

- Validity of a piston flow model: it considers an air-water interface perpendicular to the pipe direction, and its application depends mainly of the water velocity, the internal pipe diameter, and the pipe slope. Others two phases models can be used in more complex situations.

- Validity of a rigid water column model (RWCM): a RWCM could predict adequately the behavior of the water column during all experiments in this research. However, the application of an elastic water model (EWM) can be used to simulate emptying processes, which can give more information when using more complex hydraulic systems.

- Analysis in water distribution systems: this doctoral book focused on the analysis of emptying processes of a single pipe and a pipeline of undulating profile. However, the analysis in water distribution systems was not studied. Then, researches can be conducted with new installations and layouts.

- The backflow air phenomenon cannot be simulated with the mathematical model since the admitted air, when draining the valves, needs to be addressed with additional formulations requiring new experiments. In this sense, a 2D CFD model could represent more accurately the backflow phenomenon, but it requires greater computing times compared to the mathematical model.

It is also important to note that in pipelines operating in real conditions is practically impossible to know the initial conditions of emptying processes with respect to air pocket sizes and their location. As a consequence, many engineers use recommendations of manuals and manufacturers to avoid the risk of pipelines collapse.

Appendix A

CFD model for predicting emptying processes of water

This appendix focuses on a comparison between the mathematical model and a 2D CFD model showing results of air pocket pressure during emptying processes of water in the two used experimental facilities (see Chapters 3 and 4). The improvement of computers and the computational methods have urged researchers to use CFD models for their numerical calculations to describe the beahvior of water column. Simulations using a 2D CFD were performed at the University of Lisbon by Besharat et al. (2018a,b).

A.1   Implementation of a 2D CFD model

Some considerations to implement a 2D CFD model were: (i) a 2D CFD simulation describes adequately experimental results; (ii) an unstructured triangular mesh was considered; (iii) a sliding mesh zone was used to compute the movement of drain valves; (iv) the explicit volume of fluid (VOF) multiphase model was considered using the enhanced wall treatment (EWT) feature for near-wall calculations; (v) the pressure implicit with the splitting of operators (PISO) method for coupling the pressure-velocity formulations was used; (vi) the spatial discretization was performed using a finite volume; (vii) the pressure staggering option (PRESTO) discretization scheme is used for the pressure; (vii) drain valves actuation was conducted using a user defined function (UDF) for the different opening percentages and valve maneuvering times; and (vii) a time step of 0,001 s was used for all simulations. A complete description of the 2D CFD simulations is presented by Besharat et al. (2018a,b). For both pipe installations simulations were executed on a desktop computer: Intel(R) Core(TM) i7 − 4790K CPU @ 4.00 GHz, 4 Core(s), 8 Logical Processor(s) with an installed physical memory of 16 GB. The ANSYS Fluent R18.2 academic version was used to run simulations.

The two-equation $k - \varepsilon$ was selected to perform simulations since it is capable to fix with a great accuracy experiments of air pocket pressure consuming low computing times. CFD simulations were conducted using formulations as follows:

$$\frac{\partial}{\partial t}(\rho_m) + \nabla(\rho_m \vec{v}) = 0 \tag{A.1}$$

$$\frac{\partial}{\partial t}(\rho_m \vec{v}) + \nabla.(\rho_m \vec{v}\vec{v}) = -\nabla p + \nabla.[\mu_m(\nabla \vec{v} + \vec{v}^{-T})] + \rho_m \vec{g} + \vec{F} \tag{A.2}$$

where $\vec{v}$ is the fluid velocity, p is the static pressure, $\vec{g}$ is the gravitational acceleration, $\vec{F}$ is the body force, ρ_m is the mixture density and μ_m is the mixture viscosity. The ρ_m and μ_m are functions of the phase volume fraction and density/viscosity of each phase as defined below:

$$\rho_m = \alpha_a \rho_a + (1 - \alpha_a)\rho_l \tag{A.3}$$

$$\mu_m = \alpha_a \mu_a + (1 - \alpha_a)\mu_l \tag{A.4}$$

where α_a the is the volume fraction of the air in percentage. For each mesh, if the whole cell is occupied by air, the $\alpha_a = 1$ and if there is no air in the cell $\alpha_a = 0$. The subscripts m, a and l correspond to the mixture, air, and liquid phases, respectively.

The definition of a fine mesh and the acting of drain valves were important to conduct these simulations. A structured quadrilateral mesh was used in the near-wall zone and the mesh at the inner region of the pipe was constructed by unstructured triangular cells for the single pipeline (see Chapter 3) and the pipeline of irregular profile (see Chapter 4). Table A.1 shows the number of used cells.

Table A.1: Mesh of pipeline installations

Pipeline installation	Elements	Average face size
Single pipeline (see Chapter 3)	60630	0,003
Pipeline of irregular profile (see Chapter 4)	108916	0,00305

For both pipeline installations a sliding mesh zone was implemented to represent the behavior of drain valves using a user defined function (UDF) for the different opening percentages and valves maneuvering times. Opening percentages between 6% and 12% were considered for the single pipeline; while, values between 6% and a full opening were simulated for the pipeline of irregular profile. Figure A.1 presents the generated mesh for the two used installations.

A.2 Comparison of the proposed model and a 2D CFD model

Figures A.2 and A.3 show results concerning to air pocket pressure patterns for the single pipeline and the pipeline of irregular profile, respectively. The comparison was conducted

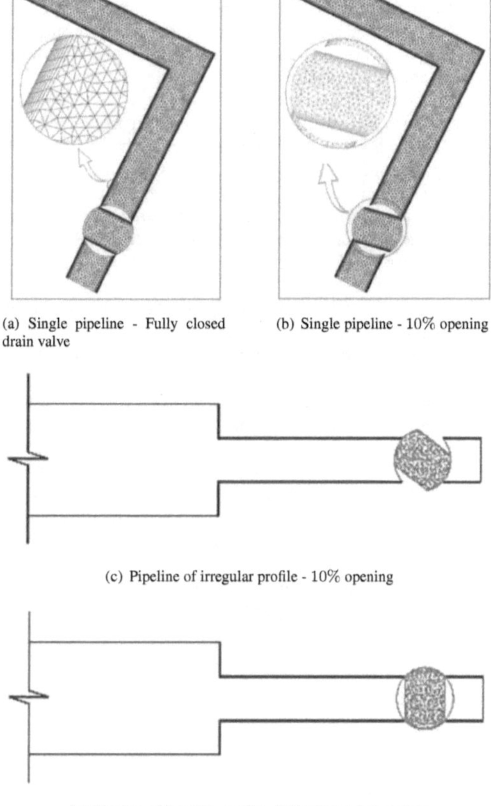

(a) Single pipeline - Fully closed drain valve

(b) Single pipeline - 10% opening

(c) Pipeline of irregular profile - 10% opening

(d) Pipeline of irregular profile - Fully closed drain valves

Figure A.1: The generated mesh in the pipe systems near drain valves

only for 3 tests (runs), which represent the majority of scenarios. The situation without admitted air during the emptying process in pressurized pipelines was performed using a 2D CFD model since it represents the more risky condition and a limitation of the proposed model. The minimum values of subatmospheric pressure are reached in the single pipeline since it presents a greater difference elevation (Δz) compared to the pipeline of irregular profiles. In all cases, both models are capable to predict adequately the majority of experiments. Three tests were selected to compare the mathematical model vs. a 2D CFD since they represent all situations. Figure A.2 shows how the mathematical model gives similar results compared to a 2D CFD model. The complexity of the pipeline of irregular profile is greater with initial air pockets distributed non-uniformly (see Figure A.3c), where the proposed model can only represent the first oscillation and the drop of subatmospheric pressure. This situation can be represented by the 2D CFD model. Air pockets distributed non-uniformly generate a rapid movement of the air intrusion by drain valves, which are not modeled by the proposed model. Simulation times using the proposed model take around 5 s, however using a 2D CFD model take around 8 h.

In all cases, the proposed model can be only applied during the first seconds of the transient flow, since after that the air starts to come into pipe systems by drain valves. This phenomenon is known as backflow air. The understanding of the backflow air occurrence was developed using the 2D CFD model. Figure A.4 shows the backflow air occurrence for Test No. 1 in the single pipeline where a small quantity of air starts to inject into the pipe at 1,0 s. The phenomenon relieves subatmospheric absolute pressure since the injected air is at atmospheric conditions. In this sense, the mathematical model is a reliable tool since it brings higher values of subatmospheric pressure pattern compared to experiments after first seconds of the beginning of the hydraulic event. When air valves are acting in hydraulic installations, the backflow air phenomenon is not presented since the water is coming out rapidly by drain valves. As a consequence, simulations using a 2D CFD model using air valves were not performed.

In summary, the proposed model and the 2D CFD model can be used to represent the emptying process of water in pressurized pipelines.

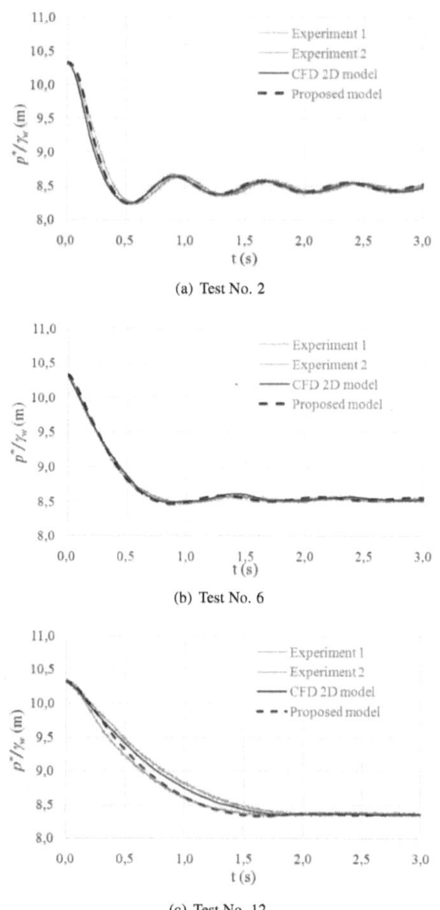

(a) Test No. 2

(b) Test No. 6

(c) Test No. 12

Figure A.2: Comparison between the mathematical model and the 2D CFD model of air pocket pressure for the single pipeline located at the Universitat Politècnica de València (see Chapter 3)

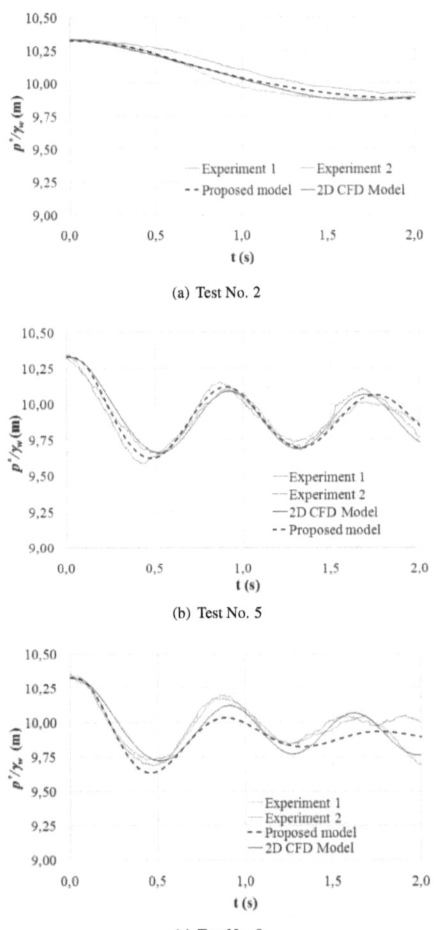

(a) Test No. 2

(b) Test No. 5

(c) Test No. 8

Figure A.3: Comparison between the mathematical model and the 2D CFD model of air pocket pressure for the pipeline od

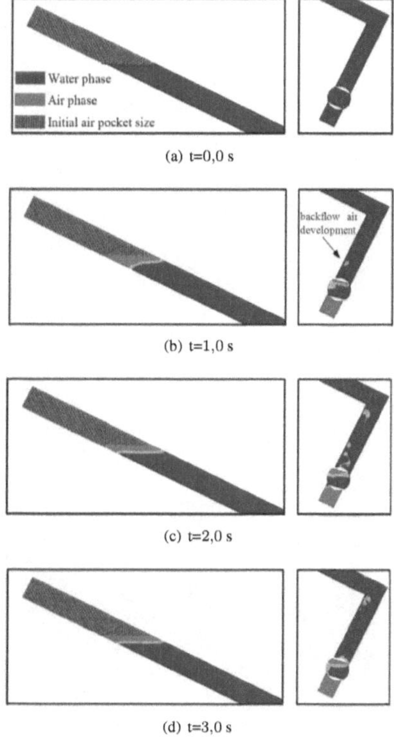

(a) t=0,0 s

(b) t=1,0 s

(c) t=2,0 s

(d) t=3,0 s

Figure A.4: Backflow air development for Test No. 1 in the single pipeline